Supporting Research and Data Analysis in NASA's Science Programs

Engines for Innovation and Synthesis

Task Group on Research and Analysis Programs
Space Studies Board
Commission on Physical Sciences, Mathematics, and Applications
National Research Council

NATIONAL ACADEMY PRESS
Washington, D.C. 1998

NOTICE: The project that is the subject of this report was approved by the Governing Board of the National Research Council, whose members are drawn from the councils of the National Academy of Sciences, the National Academy of Engineering, and the Institute of Medicine. The members of the task group responsible for the report were chosen for their special competences and with regard for appropriate balance.

The National Academy of Sciences is a private, nonprofit, self-perpetuating society of distinguished scholars engaged in scientific and engineering research, dedicated to the furtherance of science and technology and to their use for the general welfare. Upon the authority of the charter granted to it by the Congress in 1863, the Academy has a mandate that requires it to advise the federal government on scientific and technical matters. Dr. Bruce Alberts is president of the National Academy of Sciences.

The National Academy of Engineering was established in 1964, under the charter of the National Academy of Sciences, as a parallel organization of outstanding engineers. It is autonomous in its administration and in the selection of its members, sharing with the National Academy of Sciences the responsibility for advising the federal government. The National Academy of Engineering also sponsors engineering programs aimed at meeting national needs, encourages education and research, and recognizes the superior achievements of engineers. Dr. William A. Wulf is president of the National Academy of Engineering.

The Institute of Medicine was established in 1970 by the National Academy of Sciences to secure the services of eminent members of appropriate professions in the examination of policy matters pertaining to the health of the public. The Institute acts under the responsibility given to the National Academy of Sciences by its congressional charter to be an adviser to the federal government and, upon its own initiative, to identify issues of medical care, research, and education. Dr. Kenneth I. Shine is president of the Institute of Medicine.

The National Research Council was organized by the National Academy of Sciences in 1916 to associate the broad community of science and technology with the Academy's purposes of furthering knowledge and advising the federal government. Functioning in accordance with general policies determined by the Academy, the Council has become the principal operating agency of both the National Academy of Sciences and the National Academy of Engineering in providing services to the government, the public, and the scientific and engineering communities. The Council is administered jointly by both Academies and the Institute of Medicine. Dr. Bruce Alberts and Dr. William A. Wulf are chairman and vice chairman, respectively, of the National Research Council.

Support for this project was provided by Contract NASW 96013 between the National Academy of Sciences and the National Aeronautics and Space Administration. Any opinions, findings, conclusions, or recommendations expressed in this report are those of the author(s) and do not necessarily reflect the view of the organizations or agencies that provided support for this project.

International Standard Book Number 0-309-06275-6

Copyright 1998 by the National Academy of Sciences. All rights reserved.

Copies of this report are available from:

Space Studies Board
National Research Council
2101 Constitution Avenue, N.W.
Washington, D.C. 20418

Printed in the United States of America

TASK GROUP ON RESEARCH AND ANALYSIS PROGRAMS

ANTHONY W. ENGLAND, University of Michigan, *Chair*
JAMES G. ANDERSON, Harvard University
MAGNUS HÖÖK, Texas A&M University
JURI MATISOO, IBM Research (retired)
ROBERTA BALSTAD MILLER, CIESIN–Columbia University
DOUGLAS D. OSHEROFF, Stanford University
CHRISTOPHER T. RUSSELL, University of California at Los Angeles
STEVEN W. SQUYRES, Cornell University
PAUL G. STEFFES, Georgia Institute of Technology
JUNE M. THORMODSGARD, U.S. Geological Survey
EUGENE H. TRINH, Jet Propulsion Laboratory
ARTHUR B.C. WALKER, JR., Stanford University
PATRICK JOHN WEBBER, Michigan State University

PAMELA L. WHITNEY, Study Director
RONALD M. KONKEL, Consultant
ANNE K. SIMMONS, Senior Program Assistant

SPACE STUDIES BOARD

CLAUDE R. CANIZARES, Massachusetts Institute of Technology, *Chair*
MARK R. ABBOTT, Oregon State University
FRAN BAGENAL, University of Colorado at Boulder
JAMES P. BAGIAN,* Environmental Protection Agency
DANIEL N. BAKER, University of Colorado at Boulder
LAWRENCE BOGORAD,* Harvard University
DONALD E. BROWNLEE,* University of Washington
ROBERT E. CLELAND, University of Washington
JOHN J. DONEGAN,* John Donegan Associates, Inc.
GERARD W. ELVERUM, JR., TRW Space and Technology Group (retired)
ANTHONY W. ENGLAND,* University of Michigan
MARILYN L. FOGEL, Carnegie Institution of Washington
MARTIN E. GLICKSMAN,* Rensselaer Polytechnic Institute
RONALD GREELEY, Arizona State University
BILL GREEN, former member, U.S. House of Representatives
CHRISTIAN JOHANNSEN, Purdue University
ANDREW H. KNOLL, Harvard University
JANET G. LUHMANN,* University of California at Berkeley
JONATHAN I. LUNINE, University of Arizona
ROBERTA BALSTAD MILLER, CIESIN–Columbia University
BERRIEN MOORE III,* University of New Hampshire
KENNETH H. NEALSON,* University of Wisconsin at Milwaukee
GARY J. OLSEN, University of Illinois at Urbana-Champaign
MARY JANE OSBORN, University of Connecticut Health Center
SIMON OSTRACH,* Case Western Reserve University
MORTON B. PANISH,* AT&T Bell Laboratories (retired)
CARLÉ M. PIETERS,* Brown University
THOMAS A. PRINCE, California Institute of Technology
MARCIA J. RIEKE,* University of Arizona
PEDRO L. RUSTAN, JR., U.S. Air Force (retired)
JOHN A. SIMPSON,* University of Chicago
GEORGE L. SISCOE, Boston University
EUGENE B. SKOLNIKOFF, Massachusetts Institute of Technology
EDWARD M. STOLPER, California Institute of Technology
NORMAN E. THAGARD, Florida State University
ALAN M. TITLE, Lockheed Martin Advanced Technology Center
RAYMOND VISKANTA, Purdue University
PETER VOORHEES, Northwestern University
ROBERT E. WILLIAMS,* Space Telescope Science Institute
JOHN A. WOOD, Harvard-Smithsonian Center for Astrophysics

JOSEPH K. ALEXANDER, Director (as of February 17, 1998)
MARC S. ALLEN, former Director (through December 12, 1997)

---

*Former member.

COMMISSION ON PHYSICAL SCIENCES, MATHEMATICS, AND APPLICATIONS

ROBERT J. HERMANN, United Technologies Corporation, *Co-chair*
W. CARL LINEBERGER, University of Colorado, *Co-chair*
PETER M. BANKS, Environmental Research Institute of Michigan
WILLIAM BROWDER, Princeton University
LAWRENCE D. BROWN, University of Pennsylvania
RONALD G. DOUGLAS, Texas A&M University
JOHN E. ESTES, University of California at Santa Barbara
MARTHA P. HAYNES, Cornell University
L. LOUIS HEGEDUS, Elf Atochem North America, Inc.
JOHN E. HOPCROFT, Cornell University
CAROL M. JANTZEN, Westinghouse Savannah River Company
PAUL G. KAMINSKI, Technovation, Inc.
KENNETH H. KELLER, University of Minnesota
KENNETH I. KELLERMANN, National Radio Astronomy Observatory
MARGARET G. KIVELSON, University of California at Los Angeles
DANIEL KLEPPNER, Massachusetts Institute of Technology
JOHN KREICK, Sanders, a Lockheed Martin Company
MARSHA I. LESTER, University of Pennsylvania
NICHOLAS P. SAMIOS, Brookhaven National Laboratory
CHANG-LIN TIEN, University of California at Berkeley

NORMAN METZGER, Executive Director

Foreword

The charter of the National Aeronautics and Space Administration (NASA), Public Law 85-568, is known as the National Aeronautics and Space Act of 1958 with its several amendments. Title I gives a "Declaration of Policy and Purpose" listing several objectives "of the aeronautical and space activities of the United States." The first of these is "the expansion of human knowledge of the Earth and of phenomena in the atmosphere and space." It provides the rationale for most of NASA's scientific research.

One component of NASA's approach to meeting the objective of Title I is to conduct space missions. These missions consume the majority of the agency's attention and resources and are most evident to the public; they are certainly necessary for collecting the data that can drive the expansion of knowledge.

Equally important components, but ones that are generally less visible and less well appreciated, are the programs in research and analysis and in data analysis. The former provides the scientific underpinnings and often the enabling technology for NASA missions, and the latter turns their raw data into scientific understanding. Both programs are really aggregations of numerous investigations by individuals or consortia at universities, NASA centers, other federal and not-for-profit laboratories, and industry, covering a broad range of topics and kinds of activity. Each one is generally modest, but the total is a significant fraction of NASA's science expenditures.

This report takes a broad look at the research and data analysis (R&DA) programs across all the science disciplines addressed by NASA. It considers the role of R&DA, examines as much as possible the historical trends in funding, and considers ways in which R&DA programs could be improved in the context of the current space research environment.

It seems inevitable that specific space missions will continue to occupy the foreground of NASA's image, especially for those who look at the agency from some distance. Officials and policy makers, however, must give equal attention to the activities of R&DA, which are essential in meeting the agency's overarching mission to expand human knowledge.

Claude R. Canizares
Chair, Space Studies Board

Acknowledgment of Reviewers

This report has been reviewed by individuals chosen for their diverse perspectives and technical expertise, in accordance with procedures approved by the National Research Council's (NRC's) Report Review Committee. The purpose of this independent review is to provide candid and critical comments that will assist the authors and the NRC in making the published report as sound as possible and to ensure that the report meets institutional standards for objectivity, evidence, and responsiveness to the study charge. The contents of the review comments and draft manuscript remain confidential to protect the integrity of the deliberative process. We wish to thank the following individuals for their participation in the review of this report:

George Clark, Massachusetts Institute of Technology,
Arthur Code, WIYN Consortium, Inc.,
Thomas M. Donahue, University of Michigan,
Richard Goody, Harvard University (emeritus),
Jeanne Griffith, National Science Foundation,
Kenneth C. Jezek, Byrd Polar Research Center,
Adrian D. LeBlanc, Baylor College of Medicine, Methodist Hospital,
Ronald F. Probstein, Massachusetts Institute of Technology,
Roland W. Schmitt, Rensselaer Polytechnic Institute (retired), and
George Wetherill, Carnegie Institution of Washington.

We also wish to thank Kathryn Schmoll, University Corporation for Atmospheric Research, for her review comments on the data sections of the report.

Although the individuals listed above have provided many constructive comments and suggestions, responsibility for the final content of this report rests solely with the authoring committee and the NRC.

Contents

EXECUTIVE SUMMARY 1

1 INTRODUCTION 7

    1.1 Science: The Fulcrum of the Civilian Space Program, 7
    1.2 Critical Science Questions, 8
    1.3 Revitalization, 9
    1.4 Balance Between R&DA Programs and Flight Projects, 9

2 CONTRIBUTIONS OF THE RESEARCH AND DATA ANALYSIS PROGRAMS 11

    2.1 Discoveries That Influence Societal and Economic Issues and Policies, 12
    2.2 Breakthroughs That Change Scientific Understanding, 16
    2.3 Technologies That Enable New Observations, 21
    2.4 Information That Improves Mission Design, 24
    2.5 Investments That Increase the Productivity of Flight Projects, 26
    2.6 Research That Complements the Work of Other Federal Agencies, 30
    2.7 Science-driven Adventure That Stimulates Interest in Math, Science, or Engineering Education, 31
    2.8 Summary Comments, 33

3 THE ROLE OF THE RESEARCH AND DATA ANALYSIS PROGRAMS 34

    3.1 Understanding the Basis of R&DA, 34
    3.2 Understanding the Roles of R&DA, 37
    3.3 Posing a Strategy for R&DA Programs, 42
    3.4 Responding to the Changing Environment, 43

4 BUDGET TRENDS FOR THE RESEARCH AND DATA ANALYSIS PROGRAMS   45

    4.1   Overall NASA Funding Trends for R&DA: FY 1991-1998, 46
    4.2   Distribution by Sector of NASA Funding for Basic Research: FY 1991-1997, 51
    4.3   University Grants and Contracts by Type of Activity: FY 1986-1995, 51
    4.4   University Grants and Contracts: Award Sizes and Durations, 54
    4.5   Characteristics of Grants at NASA Field Centers, 56

5 SCIENCE COMMUNITY'S PERCEPTIONS ABOUT THE RESEARCH AND DATA ANALYSIS PROGRAMS   58

    5.1   Resource Allocation, 58
    5.2   Technology, Facilities, and Infrastructure, 60
    5.3   Research Grant Management, 61
    5.4   Intellectual Capital, 61
    5.5   Other Perceptions, 62

6 FINDINGS AND RECOMMENDATIONS   63

    6.1   Principles for Strategic Planning, 63
    6.2   Innovation and Infrastructure, 64
    6.3   Management of the Research and Data Analysis Programs, 65
    6.4   Participation in the Research and Data Analysis Programs, 66
    6.5   Creation of Intellectual Capital, 67
    6.6   Accounting as a Management Tool in the Research and Data Analysis Programs, 67

APPENDIXES

A	Sources of Data and Method of Development	71
B	Overview of NASA Structure and Budget	90
C	Acronyms and Abbreviations	93
D	Biographical Information for Task Group Members	96

Executive Summary

Effective science, clearly a mandate for the National Aeronautics and Space Administration (NASA), involves asking significant questions about the physical and biological world and seeking definitive answers. Its product is new knowledge that has value to the nation. NASA's flight projects are highly visible and usually the most costly elements of this process, but they are only a part of the science enterprise. Flight projects are founded on research that defines clear scientific goals and questions, designs missions to address those questions, and develops the required technologies to accomplish the missions. This research is funded primarily by NASA's research and analysis (R&A) programs. Data from flight projects are transformed into knowledge through analysis and synthesis—research that is funded both by R&A and by the data analysis (DA) portion of mission operations and data analysis (MO&DA) programs. R&A and DA programs are the subject of this report and are grouped for convenience under the heading of research and data analysis (R&DA).[1]

Although there has been relatively widespread agreement about the importance of R&DA within the scientific community, senior agency managers and key decision makers outside NASA often have found the roles filled by these programs difficult to articulate and to prioritize. The diversity and "softness" of R&DA activities compared to the sharp outlines of specific spaceflight missions have made R&DA particularly vulnerable during times of constrained resources and changing institutional structure and strategy. With the emergence of NASA's emphasis on streamlining missions, accelerating development cycles, accentuating innovation, and reducing costs—the "smaller, faster, cheaper" approach—the roles of R&DA in framing scientific issues, developing the necessary new technologies for future missions, and mining the data from extant missions to produce new scientific knowledge have become even more critical.

In 1996 the Space Studies Board (SSB) formed the multidisciplinary Task Group on Research and Analysis Programs to study R&DA programs and trends in light of new agency approaches to space

---

[1] The task group originally coined the composite term "R&DA" to designate research and data analyses that were funded outside of spaceflight projects. Because NASA budgets do not separate cleanly this way, R&DA became a catch-all surrogate for all science-related activities that were funded outside of spaceflight projects. More specific alternatives to "R&DA" were defined for the discussion of budget trends in Chapter 4. See also section 3.2 in Chapter 3.

research. In creating the task group, special attention was given to involving a mix of scientists with long-standing familiarity with NASA science programs and "newcomers" who could bring a fresh perspective to the SSB's analyses. Efforts were also made to seek wide input from the research community via consultations with the SSB's discipline-specific standing committees, invitations for comments from members of key professional societies, and solicitation of comments to the task group on the Internet. The task group also engaged a consultant with expertise in the budgeting process to assist in compiling historical data on NASA science budgets for use in studying trends in resource allocations.

The statement of task for the study identified a number of areas that would be appropriate topics for review. These included evolution of the character of R&DA projects; evolution of the relative roles of universities and NASA centers in R&DA programs; the relationship between R&DA, advanced technology development, and MO&DA programs; characteristics of R&DA projects judged to be successful in supporting a smaller, faster, cheaper approach to flight missions; assessment of the expectations for R&DA in different NASA science offices; management issues for R&DA; and options for strengthening the program in the current NASA environment. These areas provided general guideposts at the beginning of the study; specific topics emerged during the review to become focal points for attention.

SCOPE AND CONTENT OF REPORT

Chapter 1 of this report provides an introduction to the role and character of projects included in R&DA and summarizes the motivation for the study. Chapter 2 focuses on questions of the actual breadth and depth of impact of R&DA programs. In reviewing the history of research conducted under R&DA in NASA's three science offices—space science, Earth science, and life and microgravity science—the task group developed a sampler of specific accomplishments that illustrate the return on investments in R&DA. These examples highlight seven different kinds of contributions, namely:

1. Discoveries that influence societal and economic issues and policies;
2. Breakthroughs that change scientific understanding;
3. Technologies that enable new observations;
4. Information that improves mission design;
5. Investments that increase the productivity of flight projects;
6. Research that complements the work of other federal agencies; and
7. Science-driven adventure that stimulates interest in math, science, or engineering education.

Although the treatment of R&DA in different NASA offices often has been fragmented and nonuniform, the task group adopted (Chapter 3) a set of seven elements that form a suitable organizing framework:

1. Theoretical investigations;
2. New instrument development;
3. Exploratory or supporting ground-based and suborbital research;
4. Interpretation of data from individual or multiple space missions;
5. Management of data;
6. Support of U.S. investigators who participate in international missions; and
7. Education, outreach, and public information.

A fundamental premise of this study is that these seven activities are integral elements of an effective research program strategy; thus, they must be explicitly linked to the strategic plan of the science organization.

The task group's analysis of NASA budget data (Chapter 4) focuses on four areas:

1. Overall funding trends for R&DA from FY 1991 to 1998;
2. Distribution of funding for basic research among NASA laboratories, private industry, academia, and other organizations from FY 1991 to 1997;
3. Distribution of funding to universities by type of activity (e.g., research, development, operations, training) from FY 1986 to 1995; and
4. Number and size of research awards to universities from FY 1986 to 1995.

These data illuminate a number of issues regarding the balance between funding for R&DA and for flight programs and the balance between different kinds of activities within NASA's R&DA portfolio.

Chapter 5 summarizes a number of concerns and perceptions about R&DA support as viewed, often anecdotally, in the research community and notes where the task group's budget trend analysis can illuminate the concerns quantitatively. In Chapter 6, the task group's conclusions are framed in terms of a set of strategic principles, an overarching finding that emerges from the study, and a set of six recommendations to NASA regarding the management of R&DA programs in the three science offices. These six recommendations cover the following areas:

1. Principles for strategic planning,
2. Innovation and infrastructure,
3. Management of the R&DA programs,
4. Participation in the R&DA programs,
5. Creation of intellectual capital, and
6. Accounting as a management tool in the R&DA programs.

FINDINGS AND RECOMMENDATIONS

Principles for Strategic Planning

Finding: The task group finds that R&DA is not always thoroughly and explicitly integrated into the NASA enterprise strategic plans and that not all decisions about the direction of R&DA are made with a view toward achieving the goals of the strategies. The task group examined the trend and balance of R&DA budgets and found alarming results (Chapter 4, Sections 4.1 and 4.3); it questions whether these results are what NASA intends.

Recommendation 1: The task group recommends that each science program office at NASA do the following:

• Regularly evaluate the impact of R&DA on progress toward the goals of the strategic plans.
• Link NASA research announcements (NRAs) to addressing key scientific questions that can be related to the goals of these strategic plans.
• Regularly evaluate the balance between the funding allocations for flight programs and the

R&DA required to support those programs (e.g., assess whether the current program can support R&DA for the International Space Station).

• Regularly evaluate the balance among various subelements of the R&DA program (e.g., theoretical investigations; new instrument development; exploratory or supporting ground-based and suborbital research; interpretation of data from individual or multiple space missions; management of data; support of U.S. investigators who participate in international missions; and education, outreach, and public information).

• Use broadly based, independent scientific peer review panels to define suitable metrics and review the agency's internal evaluations of balance.[2]

• Examine ways to maximize familiarity with contemporary advances and directions in science and technology in the process of managing R&DA, for example, via the appropriate use of rotators.[3]

Innovation and Infrastructure

Finding: Although there are sporadic funding opportunities for research infrastructure, there is no systematic assessment of the state of the research infrastructure, nor are there coherent programs to address weaknesses in the infrastructure base (Section 5.2).

Recommendation 2: The task group recommends that NASA take the following actions on research infrastructure:

• Conduct an initial assessment of the need and potential for acquiring and sustaining infrastructure in universities and field centers.

• Determine options for minimizing duplication of expensive research facilities.

• Evaluate the level of support for infrastructure in the context of the overall direction and plans for R&DA activities.

• Maximize the use of infrastructure by supporting partnering between universities and field centers.

• Explore approaches for providing peer review and oversight of infrastructure investments, which should include regular evaluation of a facility's role and contribution as a national academic resource, its degree of scientific and technical excellence, and its contribution to NASA's long-term technology planning and development.

• Institute periodic assessment of the research infrastructure in university and NASA field centers to ensure that the infrastructure is appropriate for current programs.

---

[2] National Research Council (NRC), Space Studies Board, "On NASA Field Center Science and Scientists," letter to NASA Chief Scientist France Cordova, March 29, 1995; NRC, Space Studies Board and the Committee on Space Biology and Medicine, "On Peer Review in NASA Life Sciences Programs," letter to Dr. Joan Vernikos, director of NASA's Life Sciences Division, July 26, 1995; NRC, Space Studies Board, "On the Establishment of Science Institutes," letter to NASA Chief Scientist France Cordova, August 11, 1995.

[3] Federal agencies have used rotators—scientists from outside the federal government—for 1 to 2 years to participate in management of research programs. NASA has used interagency personnel appointments—visiting scientists administered by the Universities Space Research Association and the Jet Propulsion Laboratory—as rotators to circulate new ideas and new individuals, on temporary appointments, into the agency system.

Management of the Research and Data Analysis Programs

Finding: The median size of NASA research grants to universities decreased in constant FY 1995 dollars from $64,000 per year in FY 1986 to $59,000 in FY 1995 for the Office of Space Science disciplines, remained relatively flat at $79,000 for Earth science disciplines, and grew from $69,000 to $100,000 for life and microgravity science disciplines during the period from 1986 to 1995 (Section 4.4, Figure 4.3). (These award sizes compare to a median of $85,000 at the National Science Foundation and a mean of between $110,000 and $120,000 at the Environmental Protection Agency.) It is well known that a single researcher cannot support a salary and a graduate student at grant levels of $50,000 and that such researchers must seek additional grants to maintain a viable research program.

Recommendation 3: NASA should routinely examine the size and number of grants awarded to individual investigators to ensure that grant sizes are adequate to achieve the proposed research and that their number is consistent with the time commitments of each investigator. The differences in award sizes for the Offices of Space Science, Earth Science, and Life and Microgravity Science and Applications should be reconciled with program objectives, especially those for space sciences, which often are funded at levels of less than $50,000 to $60,000. Where warranted, actions should be taken to address the deficiencies.

Participation in the Research and Data Analysis Programs

Finding: The task group recognizes that university-based instrument development projects led by principal investigators (PIs) can provide important training and versatility for graduate students in NASA-funded sciences. Often, innovative instrument prototypes can be developed at a fraction of the cost of facility instruments, and the analysis of instrument data and the preparation of high-quality scientific results are closely coupled with understanding of and experience in the design of scientific instrumentation. However, although the university arena frequently offers these opportunities, the task group also recognizes that some research facilities do not offer training advantages, that the economies of scale for some facility development projects are high, and that support of nonuniversity, multiuser facilities is sometimes necessary.

Recommendation 4: NASA should preserve a mix of PI-university awards and nonuniversity funding for the development of technologies, instruments, and facilities. NASA should make these decisions within the agency's overall plan for R&DA activities (Recommendation 1), with sensitivity to the advantages of the academic environment but guided by peer review of scientific and technical merit.

Creation of Intellectual Capital

Finding: NASA's principal graduate student fellowship programs are all tied to student research interests or concentrations.

Recommendation 5: NASA should explore using training grants like those of the National Institutes of Health and the National Science Foundation for first-year graduate students as a possible alternative to supporting these students as research assistants or NASA fellows. These training grants should be designed to ensure breadth in graduate education and thereby may expand students' opportunities for employment within or beyond NASA-funded sciences.

Accounting as a Management Tool in the Research and Data Analysis Programs

Finding: NASA does not use the extended records of its budgets and expenditures as management tools to monitor the health of its R&A and DA programs. Moreover, the fragmented budget structure for R&DA makes it difficult for the scientific community to understand the content of the program and for NASA to explain the content to federal budget decision makers.

Recommendation 6: NASA's science offices should establish a uniform procedure for tracking budgets and expenditures by the class of activities and the types of organizations (including intramural and extramural laboratories, industry, and nonprofit entities) that are actually performing the work. These data should be gathered and reported annually and used to inform regular evaluations of R&DA activities (Recommendations 1 and 2). One approach would be to itemize the following elements in the budget: theoretical investigations; new instrument development; exploratory or supporting ground-based and suborbital research; interpretation of data from individual or multiple space missions; management of data; support of U.S. investigators who participate in international missions; and education, outreach, and public information. In addition, these data should be made publicly available and reported annually to the Office of Management and Budget and to Congress.

1

Introduction

1.1 SCIENCE: THE FULCRUM OF THE CIVILIAN SPACE PROGRAM

The second half of the 20th century has been a time of momentous developments in science, medicine, and technology. The exploration and utilization of space are singular in the boldness with which they have pushed the expansion of human knowledge and engaged the imagination of citizens worldwide. In executing its programs, the National Aeronautics and Space Administration (NASA) has been carrying out the mandate expressed in the National Space Act of 1958, which emphasized the goals of expansion of knowledge, U.S. scientific and technological leadership, international cooperation, and wide dissemination of results.

In 1991, the presidentially appointed Advisory Committee on the Future of the U.S. Space Program (the "Augustine committee") ranked science first among NASA's priorities and characterized it as the program's "fulcrum."[1] The National Research Council's (NRC's) Space Studies Board (SSB) echoed a similar viewpoint a year later in its report on setting science priorities, which asserted that "development of new knowledge and enhanced understanding of the physical world and our interactions with it should be emphasized as the principal objective of space research and as a key motivation for the space program."[2]

Effective science, clearly a mandate for NASA, involves asking significant questions about the physical world and seeking definitive answers. Its product is new knowledge, and new uses of knowledge, that have value to the nation. NASA's flight projects are highly visible and usually the most costly

---

[1] *Report of the Advisory Committee on the Future of the U.S. Space Program*, U.S. Government Printing Office, Washington, D.C., Decmember 1990; National Science and Technology Council, "National Space Policy," The White House, September 1996.

[2] National Research Council, Space Studies Board, *Setting Priorities for Space Research—Opportunities and Imperatives*, National Academy Press, Washington, D.C., 1992, p. 8.

elements of this process, but they are only a part of the science enterprise. Flight projects are founded on research that defines clear scientific goals and questions, designs missions to address these questions, and develops the required technologies to accomplish the missions. This research is funded primarily by NASA's research and analysis (R&A) programs. Data from flight projects are transformed into knowledge through analysis and synthesis—research that is funded both by R&A programs and by the data analysis (DA) portion of mission operations and data analysis (MO&DA) programs. R&A and DA programs are the subject of this report and are grouped for convenience under the single heading of research and data analysis (R&DA).[3]

Beyond NASA's mandate for science, the agency embraces as an integral part of its objectives sophisticated technology—such as new microelectronics and detectors, innovative launch systems, robotics, and artificial intelligence—that enables flight projects and contributes to other terrestrial applications as well. R&DA defines, focuses, and integrates scientific and technical objectives to take maximum advantage of this technological progress. Similarly, R&DA contributes to the scientific foundation that underlies the development of many of the applications that result from NASA's work. Recent advances in such areas as global communications, navigation, and weather prediction are dependent on a combination of advanced science and advanced technology.

In speaking of "science" in NASA's programs and of the impact of R&DA activities on science, the task group means much more than just the pursuit of knowledge for its own sake. Important basic research is often stimulated by some societal need, and conversely good basic research often opens the way for new tools and approaches that are translated into societal or economic benefits. In a paper marking the fiftieth anniversary of Vannevar Bush's report *Science: The Endless Frontier*,[4] Princeton University political scientist Donald Stokes referrred to this aspect of science as "use-inspired basic research."[5] A substantial number of the key scientific questions framing NASA's science programs, especially in the Earth and life and microgravity sciences, reflect such a use-inspired orientation.

1.2 CRITICAL SCIENCE QUESTIONS

The R&DA grants and contracts that fund university and industry researchers facilitate the agency's link to the nation's intellectual resources. Through them, NASA provides science to inform public policy debate, opportunities to train the nation's young scientists and engineers, scientific developments that stimulate technology breakthroughs, and new avenues for education at all levels. These ground-based programs identify the critical science questions that can be addressed through the use of aeronautics and space technologies and through access to unique suborbital and orbital laboratories and spacecraft or deep-space probes. Among these questions are the following:

- How did the universe begin and what is its ultimate fate?
- How and where did life begin?
- How do galaxies, stars, and planetary systems form and evolve?

---

[3]The task group originally coined the composite term "R&DA" to designate research and data analyses that were funded outside of spaceflight projects. Because NASA budgets do not separate cleanly this way, R&DA became a catch-all surrogate for all science-related activities that were funded outside of spaceflight projects. More specific alternatives to "R&DA" were defined for the discussion of budget trends in Chapter 4.

[4]Vannevar Bush, *Science: The Endless Frontier*, Appendix 3, "Report of the Committee on Science and the Public Welfare," U.S. Government Printing Office, Washington, D.C., 1945.

[5]Donald Stokes, *Vannevar Bush II: Science for the 21st Century*, Sigma Xi, Research Triangle Park, N.C., 1995, p. 28.

- How is the evolution of life linked to planetary evolution and to cosmic phenomena?
- Can climate be predicted a year or a season in advance?
- How do terrestrial ecosystems respond to land cover and land use change?
- What is the role of gravity in the biological processes of plants and animals?
- How does gravity affect the common processes found in natural and industrial activities?

1.3 REVITALIZATION

In the early 1990s, NASA faced reduced budgets at a time when many flight projects were becoming increasingly complex, lengthy, and costly. To respond to this dilemma, NASA introduced a revitalization regimen based on a new model for flight projects that has been captured in the mantra "smaller, faster, cheaper." An early advocate of the new strategy was physicist Freeman Dyson,[6] who argued that "quick is beautiful," meaning that "smaller and less cumbersome space-science missions" could more easily respond to new ideas. Smaller, faster, cheaper missions promise a robust flight rate by scaling many flight projects within the science enterprise to smaller launch vehicles that cost tens of million dollars rather than the hundreds of million dollars of larger vehicles; to durations of 3 to 6 years rather than the career-consuming 10 to 15 years that were becoming common; and to project costs that permit risk taking rather than the billion-dollar costs that freeze out innovation. Under the new model, costs are contained through small payloads, sharply reduced project lifetimes, an emphasis on technical innovation, and a willingness to compromise the breadth of science objectives to achieve more limited objectives more quickly. Although larger missions will continue to be important to achieve some science objectives,[7] smaller missions are now a significant portion of NASA's program.

As the faster-paced style of the agency has begun to take hold, some concerns are also emerging. At some levels of NASA management, the central and unique nature of R&DA in the NASA mission is becoming blurred or even lost. Many of these perceptions have been expressed. They include the lack of clear and consistent representation of R&DA's role in congressional testimony regarding agency resources, the shifting of resources from R&DA to cover overruns in flight programs, and the transfer of responsibility from flight missions to R&DA for analysis of core data from a given mission without a commensurate transfer of funds.

1.4 BALANCE BETWEEN R&DA PROGRAMS AND FLIGHT PROJECTS

If both R&DA programs and flight projects are essential for the effectiveness of NASA's unique science, then there must be an optimum balance between them. Too few flight opportunities would cause NASA to lose the unique opportunities that research in and from space brings to science; too little investment in R&DA programs would cause NASA to lose the intellectual content needed to identify and answer significant scientific questions.

The overarching theme of this report is that the effectiveness of NASA's unique science enterprise derives from an essential balance between R&DA programs and flight projects. The task group illustrates the central role of R&DA programs through examples of exceptional successes (Chapter 2);

---

[6]Freeman Dyson, "Quick is Beautiful" and "Science and Space," pp. 135-179 in *Infinite in All Directions*, Harper & Row, New York, 1988.

[7]National Research Council, Space Studies Board, *The Role of Small Missions in Lunar and Planetary Exploration*, National Academy Press, Washington, D.C., 1995; National Research Council, Space Studies Board, *The Role of Small Satellites in NASA and NOAA Earth Observation Programs*, National Academy Press, Washington, D.C. (in preparation).

develops the tie between R&DA investments and agency strategic goals (Chapter 3); marshals available data that describe current R&DA investments and trends (Chapter 4); explores some specific concerns of scientists vis-à-vis the R&DA programs (Chapter 5); and presents its findings and recommendations (Chapter 6). Many of the points raised are not new. They have been addressed before by the Space Studies Board[8] and by other advisory bodies.[9] Finally, the task group emphasizes that this report is not an appeal for an increase in NASA funding. It is an assertion that the activities funded by R&DA programs are essential to NASA's science enterprise, that the recent decreases in R&DA funds as a fraction of NASA's science budget are harmful to the quality and productivity of NASA's investment in science, and that R&DA programs have to be integrated more consciously into NASA's strategic management practices.

---

[8]National Research Council, Space Studies Board, *Managing the Space Sciences*, National Academy Press, Washington, D.C., 1995; letter report sent by Space Studies Board Chair Claude Canizares to NASA Associate Administrator for Space Science Wesley Huntress, "On NASA's Office of Space Science draft strategic plan," August 27, 1997.

[9]Space and Earth Sciences Advisory Committee of the NASA Advisory Council, *The Crisis in Space and Earth Sciences*, November 1986; National Commission on Space, *Pioneering the Space Frontier*, Bantam Books, New York, May 1986; Steven Wofsy, *Report of the NASA Earth System Science and Applications Advisory Committee* (ESSAAC), February 12, 1997.

2

Contributions of the Research and Data Analysis Programs

The R&DA programs encompass a wide variety of activities conducted by thousands of researchers within NASA, other government laboratories, universities, and industry. Although the total budget allocated to R&DA exceeds $1.5 billion (FY 1997) in constant FY 1995 dollars, much of this consists of individual awards of $100,000 or less (see Chapter 4). Most R&DA projects originate in the creative imaginations of investigators who submit proposals within the context of a specific program in one of NASA's larger science areas—astronomy and astrophysics, solar-terrestrial interactions, planetary sciences, Earth sciences, and life and microgravity sciences. These proposals are ranked through peer review processes that assess their scientific merit, feasibility, and value to the specific program. The outcomes of R&DA activities are equally diverse, ranging from published papers to designs for new instrumentation, the award of a Ph.D. degree, and educational products for elementary schools.

This breadth and diversity, often considered one of the positive attributes of R&DA, have also made it difficult for third parties to understand the nature of these programs and to appreciate their effectiveness and value. In an attempt to describe R&DA—its array of roles, functions, and impacts on science and society—the task group collected samples of projects from each of the major science themes and organized them according to the following categories:

1. Discoveries that influence societal and economic issues and policies;
2. Breakthroughs that change scientific understanding;
3. Technologies that enable new observations;
4. Information that improves mission design;
5. Investments that increase the productivity of flight projects;
6. Research that complements the work of other federal agencies; and
7. Science-driven adventure that stimulates interest in math, science, or engineering education.

The examples cited in these seven categories are exemplary, not necessarily because of the process by which they achieved results, but because they illustrate the value that R&DA projects can have for society, for science, and for NASA's missions. These examples are not meant to be typical of all R&DA grants, nor—for disciplines that span the interests of more than one federal agency—do they necessarily represent spheres of NASA leadership, but they do illustrate how modest R&DA investments make a significant difference.

2.1 DISCOVERIES THAT INFLUENCE SOCIETAL AND ECONOMIC ISSUES AND POLICIES

R&DA projects occasionally inform policy debates about issues of national importance or change the way we live or work. Two examples of the former and one of the latter are presented in this section. In the first, results from R&DA research have helped provoke political action; in the second, emerging technologies have stimulated R&DA research in anticipation of political interest; and in the third, products of R&DA-funded research have contributed to understanding the linkages between El Niño, the Southern Oscillation, and global weather, and consequently to developing regional crop strategies in agribusiness. Box 2.1 contains additional examples of this type.

1. Discovery and diagnosis of the antarctic ozone hole, a major, unanticipated surprise for scientists, caused significant changes in public policy. The annual cycle of ozone in the stratosphere over the Antarctic has been tracked by scientists beginning with projects that were part of the International Geophysical Year in 1957. In the late 1970s, an unexplained deficit emerged in the total ozone amount in late-winter observations. In 1985, the British Antarctic Survey reported for the first time that dramatic decreases were occurring in the ozone concentration over Halley Bay and that the degree of ozone loss was worsening as the decade progressed. Theories of the cause of this unprecedented loss were put forward by serious scientific research groups in an international effort to diagnose the reason for this alarming development. In one example, investigators applied models that had been developed under a NASA R&DA project to study the upper-atmospheric photochemistry of Venus and Mars. There were several expeditions to gather more information and, in August and September 1987, NASA contributed an R&DA-funded airborne survey. An ER-2 aircraft flew from Punta Arenas, Chile, to penetrate the region of the stratosphere where ozone was disappearing. The key results are shown in Plate 2.1.

The mission demonstrated unequivocally that ozone was destroyed by chlorine and bromine radicals. The case linking chlorofluorocarbons (CFCs), the molecules that transport chlorine to the stratosphere, to the destruction of ozone over the Antarctic rests on three discoveries from this NASA mission. The first discovery was that the continental-scale region of severe ozone depletion was isolated from the rest of the stratosphere by the polar night jet, which creates a continental-scale "containment vessel." The existence of this barrier preventing exchange is shown clearly by the high-resolution aircraft data in Figure 2.1. The second discovery was the anticorrelation between O_3 and ClO that occurs within this stratospheric containment vessel. Plate 2.1 shows that on August 23, 1987, as sunlight returned to the region, O_3 had emerged from the polar night largely unaffected. Three weeks later, on September 16, ozone had eroded sharply in the presence of high ClO concentrations within the sunlit containment vessel. The third discovery emerged from R&DA-funded laboratory studies that determined the rates of key reactions responsible for the destruction of ozone by chlorine and bromine radicals in sunlight.

When taken together, the three elements in this case—each of which has appeared in and been critiqued in the international scientific literature—provide irrefutable evidence that the dramatic reduction in stratospheric ozone over the antarctic continent would not have occurred had CFCs not been

> **Box 2.1**
> **Other Examples of Discoveries That Influence Societal
> and Economic Issues and Policies**
>
> - The United States is the world leader in providing hurricane and severe storm warnings. The National Oceanic and Atmospheric Administration, with its emphasis on operational forecasting, and NASA, with its emphasis on remote sensing science and technology, have cooperated to develop the current series of weather satellites and interpretive capabilities that produce these early warnings. NASA's R&DA programs supported the development of satellite-observing technologies, the improvement of interpretive capabilities, and the creation of the discipline now referred to as "remote sensing science."[1]
> - "Space weather" refers to the energy and density of the ionized "wind" that flows outward from the Sun and to the radiation from the Sun. Major explosive events on the surface of the Sun will cause significant increases in the strength of both the wind and the radiation field. These "storms" can damage the electronics in satellites, compromise communications channels that use radio links, and bring down large electrical power-distribution grids. Theoretical numerical models, such as the solar-wind magnetosphere and magnetosphere-ionosphere-thermosphere models, have become essential elements in the National Space Weather Program and figure prominently in the creation of sophisticated forecasting tools. These models were developed with a combination of NASA R&DA and National Science Foundation funding.[2]
> - Models and modeling techniques first used to study stratospheric photochemistry some 30 years ago were applied to Venus, Mars, and the antarctic ozone hole. The success of these models in explaining the makeup of the upper atmosphere of our planetary neighbors built confidence about their use in understanding complex Earth systems.[3]
>
> ---
>
> [1]For further reading, see National Research Council, Space Studies Board, *Earth Observations from Space: History, Promise, and Reality,* National Academy Press, Washington, D.C., 1995.
> [2]For further reading, see D. Dooling, "Stormy Weather in Space," *IEEE Spectrum* 32, June 1995, pp. 64-72; "'Weather' Forecasters Work on Higher Plane," *Aviation Week and Space Technology* 143, September 18, 1995, p. 49.
> [3]National Research Council, Space Studies Board, *An Integrated Strategy for the Planetary Sciences: 1995-2010,* National Academy Press, Washington, D.C., 1994, p. 13.

synthesized and then added to the atmosphere.[1] These findings brought the industrial nations to the political consensus expressed in the Montreal Protocol of 1987. Essentially, the industrial nations agreed to stop producing chlorofluorocarbons.

2. The possibility of designing high-speed civil transport aircraft to have a negligible effect on stratospheric ozone could change future aeronautics policy. More than 20 years after the first supersonic transports began passenger service, manufacturers of large commercial aircraft are again exploring the economic viability of a Mach 2.4 high-speed civil transport (HSCT). Some of the risk associated with developing the HSCT lies in the potential destruction of stratospheric ozone by fleets of these

---

[1]For further reading, see J.G. Anderson, D.W. Toohey, and W.H. Brune, "Free Radicals Within the Antarctic Vortex: The Role of CFCs in Antarctic Ozone Loss," *Science* 251:39-46, 1991; P.O. Wennberg et al., "Removal of Stratospheric O_3 by Radicals: In Situ Measurements of OH, HO_2, NO, NO_2, ClO, and BrO," *Science* 266:398-404, 1994.

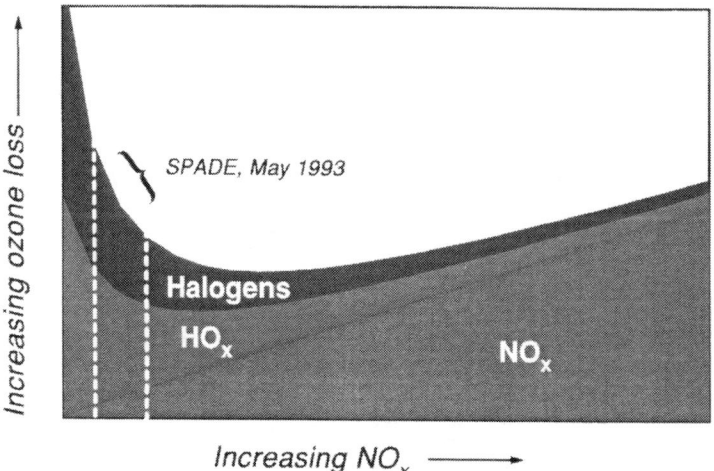

FIGURE 2.1 Removal rate of O_3 versus NO_x concentration. Because of the coupling that exists between radical families, the response of the total O_3 removal rate to changes in NO_x concentration is highly nonlinear. At sufficiently low NO_x levels, such as observed during the NASA mission, the removal rates are inversely correlated with NO_x concentration. SOURCE: P.O. Wennberg, R.C. Cohen, R.M. Stimpfle, J.P. Koplow, J.G. Anderson, R.J. Salawitch, D.W. Fahey, E.L. Woodbridge, E.R. Keim, R.S. Gao, C.R. Webster, R.D. May, D.W. Toohey, L.M. Avallone, M.H. Proffitt, M. Loewenstein, J.R. Podolske, K.R. Chain, and S.C. Wofsy, "Removal of Stratospheric O_3 by Radicals: In Situ Measurements of OH, HO_2, NO, NO_2, ClO, and BrO," *Science* 266:398-404, 1994.

aircraft. The civil transport industry in the United States will attempt to mitigate this risk by asking the federal government to resolve the environmental uncertainties associated with the HSCT. Anticipating this request, NASA developed an R&DA program to examine the response of ozone to injections of HSCT combustion products.

A fundamental premise of the perceived environmental threat from HSCTs has been that the rate of ozone removal in the stratosphere is limited by the NO_2 free radical and that a significant fleet of supersonic transports would add appreciably to the concentrations of NO_x ($x = 1,2,3$) in the lower stratosphere. R&DA projects have made two key discoveries that challenge the universality of this premise.

The first discovery emerged from an ER-2 aircraft mission over the Arctic, which demonstrated that aerosols (minute liquid droplets) have a dramatic impact on the fraction of reactive nitrogen tied up as free radicals (i.e., as NO or NO_2) in the lower stratosphere. These free radicals were being converted through a catalytic process involving the aerosols to a less-reactive nitrogen oxide, thereby providing a natural "sink" for reactive nitrogen compounds such as those found in the combustion effluent of the proposed HSCT.[2]

---

[2] D.W. Fahey, S.R. Kawa, et al., "In Situ Measurements Constraining the Role of Sulphate Aerosols in Mid-Latitude Ozone Depletion," *Nature* 363(6429):509-514, June 10, 1993. Additional information on the Airborne Arctic Stratospheric Expedition (AASE) can be found on the official NASA Web page for the AASE experiment, which is maintained by the Earth Science Division Project Office at NASA Ames Research Center, at <http://cloud1.arc.nasa.gov/aase/index.html>. In addition, the results of the AASE were published in a special issue of *Geophysical Research Letters* (see vol. 17, no. 4, 1990).

The second discovery emerged from the NASA Stratospheric Photochemistry, Aerosol and Dynamics Expedition (SPADE) mission in May 1993. Data from this ER-2 mission demonstrated that under certain conditions, the rate of catalytic ozone destruction is *inversely* correlated with total NO_x loading.[3] The relationship between ozone loss rates and added NO_x is clearly captured in Figure 2.1.

These two discoveries changed the scientific community's judgment with respect to the expected impact of the NO_x component of the HSCT effluent. Specifically, if there is a region of the stratosphere in which the addition of NO_x would actually *decrease* the rate of ozone catalytic destruction, it becomes plausible, contingent on the aircraft design and the dynamic and chemical characteristics of the stratosphere at higher altitudes, that the addition of NO_x to the lower stratosphere could leave the ozone column virtually unaffected.

3. *New understanding of the linkage between El Niño-Southern Oscillation (ENSO) and seasonal-to-interannual climate influences farming and business strategies.* Although NASA has not been a principal contributor in the investigation of ENSO, as have the National Science Foundation (NSF) and the National Oceanic and Atmospheric Administration (NOAA), the exciting advances in understanding ENSO illustrate how well-conceived R&DA projects enhance capabilities to address global-scale problems.

The occurrence of large year-to-year variations in the weather patterns of Earth is well known to everyone. They are a motive force behind major disruptions in agricultural output, fishery yields, industrial production, tourism, and international trade balances. In the western tropical Pacific, the sea surface is warm, there is extensive precipitation, and the pressure at sea level is low. In contrast, the eastern tropical Pacific is characterized by cool surface waters, limited precipitation, and high sea-level pressure. The phenomenon labeled El Niño manifests as an episodic warming of the eastern tropical Pacific. Occasionally, the warm pool in the western tropical Pacific migrates eastward, and the rainy low-pressure domain moves with it. This El Niño typically lasts three or more seasons.

The Southern Oscillation is an interannual oscillation in sea level atmospheric pressure between a region over northern Australia and a site in the central Pacific. The El Niño and the Southern Oscillation are correlated in time. The occurrence of this tropical ocean warming is correlated with dramatic changes in the fishing yields off the west coast of Peru, the character and intensity of storms along the west coast of the United States and Mexico, the propensity for hurricanes, the timing and strength of monsoons, the amount and timing of rainfall in the eastern part of Africa, and crop yields in the semiarid region of northeastern Brazil.

By the early 1980s, the link between the occurrence of ENSO and the weather patterns of the Northern Hemisphere was beginning to emerge, as were discriminating explanations of the sequence of events during oscillations between the warm and cold phases of ENSO. The massive El Niño of 1982 catalyzed scientists in two ways: (1) it irreversibly joined the oceanic and atmospheric communities in the research effort and (2) it focused the character of the scientific question and forced a level of scientific scrutiny and vitality that had not been possible before.

NASA, through its R&DA programs, provided many of the sensor technologies and satellite observing strategies that were used by NOAA, NSF, and others to identify the linkages between ENSO events

---

[3]P.O. Wennberg et al., "Removal of Stratospheric O_3 by Radicals: In Situ Measurements of OH, HO_2, NO, NO_2, ClO, and BrO," *Science* 266:398-404, 1994. Additional information on SPADE can be found on the official NASA Web page for the experiment, which is maintained by the Earth Science Division Project Office at NASA Ames Research Center, at <http://cloud1.arc.nasa.gov/spade.index.html>. In addition, results from the SPADE experiment were published in a special issue of *Geophysical Research Letters* (see vol. 21, no. 23, November 15, 1994).

and North American weather. Satellite observations such as these, when assimilated within predictive models by NOAA's National Centers for Environmental Prediction (NCEP),[4,5] are emerging as some of the most powerful examples of developing predictive ability through the balanced application of observation and theory.

2.2 BREAKTHROUGHS THAT CHANGE SCIENTIFIC UNDERSTANDING

Some scientific research alters paradigms that have been long assumed—sometimes affecting our understanding of our place in the universe. The following four examples belong to this class. Box 2.2 gives additional examples of breakthroughs in this area.

1. Chemical, structural, and isotopic analyses of meteorites may indicate past life on Mars. Among the most fundamental scientific questions is whether or not life exists or existed in the past on other planets or moons within our solar system. R&DA programs provide the primary funding for research that will eventually answer this question. The likely approach to a definitive answer will be extensive laboratory analyses of samples that have been returned from the planets and their moons, but until our technologies allow low-cost sample return or elaborate in situ analyses, the alternative is to examine meteorites found on Earth that have their origin on other planets. A number of meteorites have been identified as martian, based on their unique isotopic compositions that have been observed only on Mars's surface. Analyses of the structure and composition of these meteorites tell us something about Mars's surface at the time the material was ejected.

Two techniques have proven to be particularly powerful. The first, high-resolution transmission electron microscopy (HRTEM), reveals the structure and shape of very small inclusions within a meteorite; the second, microprobe two-step laser mass spectrometry, probes these inclusions for their composition, with a special focus on hydrocarbons. These were among the techniques recently used to find the carbonate globules and identify the polycyclic aromatic hydrocarbons (PAHs) in the Mars meteorite ALH84001 (Figure 2.2). The carbonate globules and PAHs have been interpreted as the fossil remains of life that existed on Mars more than 3.6 billion years ago—a monumental hypothesis.[6]

---

[4]For further reading, see R. Atlas, S.C. Bloom, R.N. Hoffman, E. Brin, J. Ardizzone, J. Terry, D. Bungato, and J.C. Jusem, "Geophysical Validation of NSCAT Winds Using Atmospheric Data and Analyses," *Journal of Geophysical Research,* 1998, in press; Robert Atlas, "Atmospheric Observations and Experiments to Assess Their Usefulness in Data Assimilation," *Journal of the Meteorological Society of Japan* 75 (1B):111-130, 1997; Robert Atlas, "Preliminary Evaluation of NASA Scatterometer Data and Its Application to Ocean Surface Analysis and Numerical Weather Prediction," reprinted from *Earth Observing Systems II, Proceedings of the International Society for Optical Engineering,* 28-29 July 1997, San Diego, California, Vol. 3117, pp. 90-97. Additional information can be found on the home page for the NASA Seasonal to Interannual Prediction Project (NSIPP) at <http://nsipp.gsfc.nasa.gov>.

[5]The NSIPP has been established at the Goddard Space Flight Center to develop an experimental short-term climate prediction capability for the Seasonal-to-Interannual (S-I) Climate Variability and Prediction Program in NASA's Earth Science Division. The NSIPP will serve as a central focus for this S-I program and as the primary mechanism for NASA's S-I contributions to the U.S. Global Change Research Program. The overall objective is to demonstrate the utility of satellite observations in predictions of short-term climate variations, to establish the cost-effective blend of remote surface observations and subsurface data necessary for a seasonal-to-interannual climate prediction capability, and to establish the assimilation and coupled-model systems that will provide the most reliable prediction of El Niño-Southern Oscillation (ENSO) events, other significant S-I variations, and their global teleconnections. One of the project goals is to transition the NSIPP capability and experience to the operational community, in this case the Enviromental Modeling Center of NOAA and NCEP. To date, informal connections with this group and with the Climate Prediction Center have been established.

[6]For further reading, see D.S. McKay, E.K. Gibson, Jr., K.L. Thomas-Keprta, H. Vali, C.S. Romanek, S.J. Clemett, X.D.F. Chillier, C.R. Maechling, and R.N. Zare, "Search for Past Life on Mars: Possible Relic Biogenic Activity in Martian Meteorite ALH84001," *Science* 273:924-930, 1996.

> **Box 2.2**
> **Other Examples of Breakthroughs That Change Scientific Understanding**
>
> • R&DA programs have contributed to our understanding of the causal linkage between asteroid or comet collisions with Earth and periods of dramatic climatic and biological change.[1]
> • R&DA-funded theory, models, and observations of star-forming regions of the sky have helped us understand the birth of planetary systems.[2]
>
> ---
>
> [1]National Aeronautics and Space Administration, "The Evolution of Complex and Higher Organisms" NASA SP-478, 1985; "Geological Implications of Impacts of Large Asteroids and Comets on Earth," L.T. Silver and P.H. Schultz, eds., Geological Society of America Special Paper 190, Geological Society of America, Boulder, Colo., 1982; "Global Catastrophes in Earth History; An Interdisciplinary Conference on Impact, Volcanism and Mass Mortality," V.L. Sharpton and P.D. Ward, eds., Geological Society of America Special Paper 247, Geological Society of America, Boulder, Colo., 1990.
> [2]National Research Council, Space Studies Board, *An Integrated Strategy for the Planetary Sciences 1995-2010*, National Academy Press, Washington, D.C., 1994. In addition, see *Protostars and Planets IV*, V.G. Mannings, A.P. Boxx, and S.A. Russell, eds., University of Arizona Press, Tucson, in press.

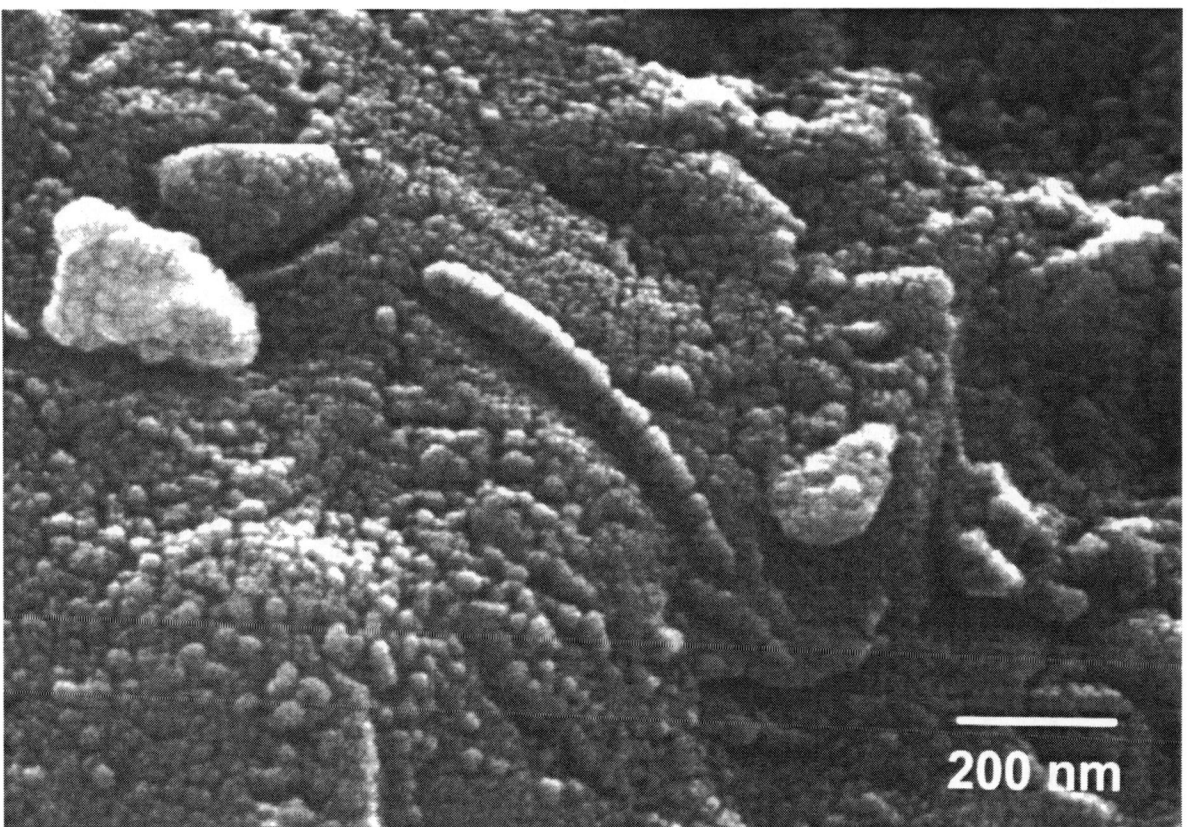

FIGURE 2.2 Mars meteorite ALH84001. Image courtesy of NASA.

Although this interpretation remains a source of debate among experts, the possibility has captured the imagination of the public and invigorated interest in exobiology within the science community. Without any doubt, this hypothesis will substantially influence plans for future Mars missions, stimulate new R&DA-funded investigations on the current catalog of meteorites, and suggest experiments to explore processes by which life itself may have been transported from one planet to another.[7]

2. *The Hubble Space Telescope data analysis has opened a new window on the universe.* The Hubble Space Telescope (HST)—developed over a period of more than two decades, launched by the shuttle in 1990, and brought into full operation at its specified spatial resolution with the refurbishment mission of 1993—has revolutionized observational astronomy with crisp images of objects ranging from protoplanetary disks and exploding stars to images of the most distant galaxies ever observed. The general public appears fascinated by the strangeness and beauty of the HST images even without understanding their content.

In late 1995, the HST director, on advice from a panel of distinguished astronomers, used his discretionary observing time to obtain the most sensitive high-resolution optical image yet obtained, the Hubble deep-field image (Plate 2.2).[8] Nearly 200 orbits of spacecraft time were used to obtain images through four filters. Perhaps the most remarkable part of the experiment was the decision to place the data in the public domain as soon as the pipeline processing was complete. Not only did this generate substantial good will among astronomers, but it also produced a host of new insights about the early epoch of galaxy evolution as diverse groups analyzed the data and shared their results rapidly—often over the Internet.

The HST Guest Observer Program is administered by the Space Telescope Science Institute at the Johns Hopkins University to support design of observing strategies, analysis of data, development of models that test hypotheses, or the publication of results. These are the intellectual elements of space-based astronomy, and they are part of an R&DA program. The HST Guest Observer Program has become one of the most productive of NASA's R&DA programs, rivaling the NSF grants program in its importance to the astronomy community. This example of world-class science based on high-quality space data conducted in full view of an interested public is an excellent model for the NASA science enterprise.[9]

3. *Recognition that microorganisms carry "molecular fingerprints" is providing a new understanding of the diversity and evolution of life on Earth.* Current excitement about the possibility of life elsewhere in the solar system has encouraged scientists to reconsider the question of how well we understand life on Earth. Recent breakthroughs in microbiology, made possible in part by support from NASA's exobiology program, are providing unexpected answers to this question—answers that necessitate a new view of evolution and ecology.

All organisms carry molecular fingerprints in the form of chemical information stored in genes. Like true fingerprints, these genetic signatures uniquely identify the species that harbor them. Unlike conventional fingerprints, however, the information in these genetic signatures allows us to reconstruct the evolutionary relationships among all living organisms—the family tree of life. Research in this area has led to the recognition of a major branch of the tree of life that no one even suspected 20 years ago—

---

[7]For further reading, see J.K. Frederickson and T.C. Onstott, "Microbes Deep Inside the Earth," *Scientific American* 275 (4, October):68-73, 1996.

[8] R.E. Williams et al., "The Hubble Deep Field: Observations, Data Reduction, and Galaxy Photometry," *Astronom. J.* 112:1335, 1996.

[9]For further reading, see D. Fisher and H. Duerbeck, *Hubble: A New Window on the Universe*, Springer-Verlag, New York, 1996.

the Archaea (Figure 2.3). Archaea look like bacteria but have different genes for managing and reading their DNA. Some Archaea are able to thrive at temperatures near the boiling point of water and at acidity levels that would etch metal. Because the Archaea branch is near the base of the tree of life, the biology of Archaea provides clues to the nature of early life on our planet. The molecular fingerprints of these and other organisms also enable microbiologists to understand for the first time the true composition of microbial communities—revealing among other things that most of the bacteria and Archaea that cycle materials in soil and water have never been cultured and remain essentially unknown.

Coupled with NASA-sponsored paleontological investigations of Earth's oldest sedimentary rocks, these studies are providing a revolutionary new perspective on early life on Earth, an understanding that will help guide future exploration of Mars and other bodies in the solar system. Because they contain enzymes capable of carrying out molecular functions under extreme conditions, Archaea and other microorganisms that thrive in unusual environments are also of growing interest in the emerging fields of biomolecular technology. Thus, these simple microorganisms hold important keys to understanding both our evolutionary past and our technological future. Research funded by the R&DA program on molecular sequences and their use in understanding the evolutionary relationships of microbial life laid the foundation for recent insights on Archaea.[10]

4. *Exposure to microgravity produces synaptic plasticity in the peripheral vestibular gravity receptors.* Human and animal responses to low-gravity environments are most noticeable in balance, orientation, and movement. These neurological processes are controlled by the vestibular system, a major component of the inner ear that helps control movement by its connections with reflex pathways in the central nervous system. Vestibular maculae on each side of the head act as gravity sensors. Research conducted on the Spacelab for Life Sciences (SLS) missions SLS-1 and SLS-2 indicated that maculae possess the property of neuronal plasticity.[11] Specifically, it was demonstrated that synapses in the hair cells of rat vestibular maculae increase significantly in number in microgravity. Other new results indicate that macular systems are sensitive to stress. The magnitude of the increments in synapses in these gravity-sensitive hair cells makes them interesting for studying the molecular basis of synaptic plasticity wherever it occurs in the nervous system. The new insights into gravity sensor capability for synaptic change should be clinically relevant to rehabilitative training and/or pharmacological treatments for vestibular disease.

Aside from having scientific and clinical value, ultrastructural findings on these sensory cells led directly to the need to visualize the complex macular microcircuitry in three dimensions, based on serial section transmission electron microscopy (TEM). This was necessary to establish the topographic relationship between nerve fibers and hair cells in the maculae and to determine precisely where synaptic changes occur in altered gravity. The ensuing work was directed toward ultrastructural analysis of synapses in hair cells exposed to microgravity and computer-based, three-dimensional visualization. The visualization software has been provided to more than 30 laboratory investigators around the country for their use. Additionally, the software produced originally for scientific research has now been adapted for medical applications. The new software has been used to produce high-fidelity three-dimensional reconstructions of the face and skull (Plate 2.3) and of the lungs and heart directly from

---

[10] For further reading, see N.R. Pace, "A Molecular View of Microbial Diversity and the Biosphere," *Science* 276:734-740, 1997.

[11] M.D. Ross, "Morphological Changes in the Rat Vestibular System Following Weightlessness," *J. Vestib. Res.* 3:241-251, 1993; M.D. Ross, "A Spaceflight Study of Synaptic Plasticity in Adult Rat Vestibular Maculas," *Acta Otolaryngol.* (Stockholm) Suppl. 516:1-14, 1994; M.D. Ross, "Synaptic Changes in Rat Maculae in Space and Medical Imaging: the Link," *Otolaryngol. Head Neck Surg.*:S25-S28, 1998.

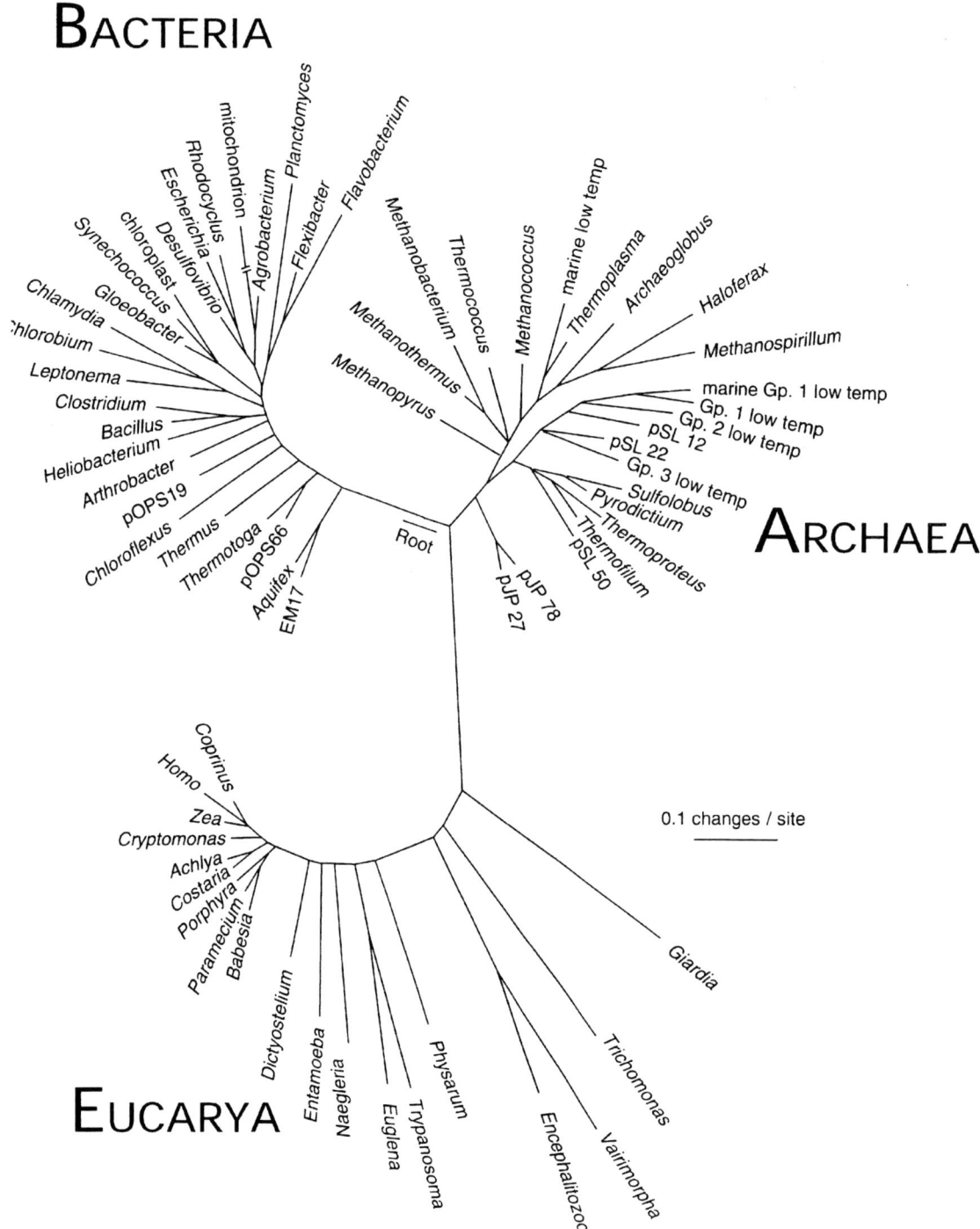

FIGURE 2.3 Universal phylogenetic tree based on small-subunit ribosomal RNA sequences. Sixty-four rRNA sequences representative of all known phylogenetic domains were aligned and a tree was produced using FASTD-NAML. That tree was modified to the composite one shown by trimming lineages and adjusting branchpoints to incorporate results of other analyses. The scale-bar corresponds to 0.1 changes per nucleotide. SOURCE: N.R. Pace, "A Molecular View of Microbial Diversity and the Biosphere," *Science* 276:734-740, 1997.

computed tomography (CT) scans of patients, and of a breast with tumor in situ from a magnetic resonance (MR) scan, among other applications.

These studies illustrate how asking simple, basic research questions can lead an investigator down paths unanticipated at the initial stage. In this example, the investigation moved from synaptic changes in gravity sensors to technology transfer by NASA that will benefit the health of the nation.

2.3 TECHNOLOGIES THAT ENABLE NEW OBSERVATIONS

NASA's goal of smaller, faster, cheaper missions means achieving the most science for the dollar. Several ways in which this might be done are discussed below. The first example describes a new technology for ground-based astronomy that was developed with R&DA funds. Critics might question whether ground-based technologies are an appropriate product for NASA. In this case, NASA had a compelling interest in the science that the new technology enabled.

The second example is one of many cases in which R&DA investments in technology have led to a new class of instruments. Although x-ray telescopes using grazing optics have become the norm, we often forget that the underlying technology has evolved since Apollo.

The third example describes the development of new technologies for flight projects with R&DA funds as a way to control mission development costs. Only large flight projects have been able to afford the time or expense of developing new technologies as part of the project. Ironically, these large projects can least afford a delay in schedule when a technology is not ready, or the loss of a mission when a new technology fails. Even after new technologies have been proven in space, managers of the more costly flight projects may be reluctant to use them because their records of successful flights are short. Additional examples of enabling technologies are given in Box 2.3.

The lower cost of smaller, faster, cheaper missions encourages project managers to take greater risks with more capable, but often less proven, technologies that are intended to compensate, to some extent, for the lower mission caps. "Faster" characteristically means 3 years between flight approval and launch rather than the 7 years that had become the norm for large flight projects. Although some

---

Box 2.3
Other Examples of Technologies That Enable New Observations

• Telescopes on the Solar and Heliospheric Observatory (SOHO), the Transition Region and Coronal Explorer (TRACE), and the proposed Solar-B missions all use optics and imaging technologies first developed under an R&DA program and then flown on suborbital platforms provided through an R&DA program.[1]

• Atmospheric UV and EUV measurements on the Upper Atmosphere Research Satellite (UARS) were made using technologies developed in the R&DA suborbital program.[2]

• R&DA-funded technology development led to the Near Infrared Camera and Multi-object Spectrometer (NICMOS), which is currently flying on the Hubble Space Telescope.[3]

---

[1] For further reading, see National Research Council, Space Studies Board, *Assessment of Programs in Solar and Space Physics*, National Academy Press, Washington, D.C., 1991.
[2] National Research Council, Space Studies Board, *An Integrated Strategy for the Planetary Sciences 1995-2010*, National Academy Press, Washington, D.C., 1994, p. 175.
[3] For further reading, see R.I. Thompson et al., "Initial On-Orbit Performance of NICMOS," *Astrophysics Journal* 492:L83, 1998.

technology development could be accomplished within the more generous development period and budget of a larger flight project, new technologies for a smaller, faster, cheaper mission must be brought to flight readiness through an R&DA program prior to initation of the flight project. The more ambitious of these new technologies might even require validation on a technology demonstration mission prior to assignment on a science mission.[12]

1. Precision Doppler measurements reveal evidence of planetary mass companions associated with other stars. If many of the hundreds of thousands of stars visible from Earth are like our Sun, what is the likelihood that any have planets orbiting them? If so, what kinds of planets might they be?

The motions of orbiting planets induce minute Doppler shifts in the emission spectra of solar-type stars. The technology needed to detect these very small Doppler shifts was, in part, a product of NASA's Innovative Research Program—an R&DA program that has been terminated.[13] Once available, astronomers quickly used the new technology to discover substellar planetary mass companions to eight solar-type stars. The rate of occurrence of planetary systems is a central issue of one of NASA's principal science themes.

2. The evolution of grazing x-ray optics has been the key enabling technology for x-ray astronomy. The disciplines of x-ray astronomy, extreme-ultraviolet (EUV) astronomy, and gamma-ray astronomy would not have developed without the first discoveries of astronomical sources by instruments on suborbital flights funded by R&DA programs. In particular, the discovery of solar x rays motivated R&DA-funded research on the x-ray optics that produced the first images of a celestial x-ray source, the Sun. The Skylab x-ray telescope, the discovery of celestial x-ray sources, and the Einstein Observatory each gave x-ray optical technologies a substantial boost, but it was the continuing R&DA support that led to the maturation of the grazing x-ray optics used, for example, in the Advanced X-ray Astrophysics Facility (AXAF).[14]

3. The Descent Imager Spectral Radiometer will measure the composition of Saturn's atmosphere. The Planetary Instrument Definition and Development Program (PIDDP) has been an R&DA-funded incubator of instruments for planetary missions. PIDDP investigators develop proof-of-concept instruments or instrument components for use in future flight projects. For example, gas-sampling instruments on Venus atmospheric probes were known to become so badly clogged by materials from Venus's clouds that they could not accurately measure atmospheric composition. To avoid the problem on future missions, PIDDP investigators developed an instrument using optical technologies that measured radiative balance, cloud density, and atmospheric composition without drawing gas samples into the probe. A laboratory proof-of-concept demonstration of the technology led to the Descent Imager Spectral Radiometer (Figure 2.4) for the Huygens entry probe on NASA's Cassini mission launched in

---

[12]NASA's New Millennium Program is a technology demonstration program to validate technology in spaceflight that will lower the risks to future science missions using the technologies. The program draws on existing government-funded research and development efforts.

[13]See G.W. Marcy, E. Williams, L. Mao, and R.P. Butler, "Precision Doppler Measurements: Detection of Other Planetary Systems," pp. 205-213 in *Remote Sensing Reviews*, Narendra S. Goel and Joseph Alexander, eds., Vol. 8, Nos. 1-3, Harwood Academic Publishers, Yverdon, Switzerland, 1993.

[14]For further reading, see R. Giaconni, N.F. Harmon, R.F. Lace, and Z. Szilagyi, *JOSA* 55, 1995, p. 345; R. Giacconi, W.P. Reidy, G.S. Vaina, L.P. Van Speybroeck, and T.F. Zehnpfenning, *Space Science Review* 9:3, 1969; B. Aschenbach, *Rep. Prog. Phys.* 48:579, 1985.

FIGURE 2.4 The Descent Imager Spectral Radiometer sensor head with front cover assembly removed to show the infrared upward- and downward-looking optics, the two-inch-diameter surface science lamp reflector, sun sensor optics, upward-looking visible spectrometer and violet detector optics, solar aureole camera, down-looking high-resolution imager optics (partially hidden by the reflector), down-looking medium-resolution imager optics, side-looking imager optics, and the various optical fibers throughout providing on-board calibration light. SOURCE: Courtesy of M.G. Tomasko, University of Arizona.

1997.[15] The Huygens probe, developed by the European Space Agency (ESA), will enter the cloudy atmosphere of Saturn's moon Titan in 2004.

---

[15]For additional information, see "Descent Imager/Spectral Radiometer Images and Spectra," prepared by M.G. Tomasko, University of Arizona, available on the European Space Agency's Huygens Probe home page at <http://www.estec.esa.nl/spdwww/huygens/html/disr.html>. Also see M.G. Tomasko, L.R. Doose, P.H. Smith, R.A. West, L.A. Soderblom, M. Combes, B. Bézard, A. Coustenis, C. deBergh, E. Lellouch, J. Rosenqvist, O. Saint-Pé, B. Schmitt, H.U. Keller, N. Thomas, and F. Gliem, "The Descent Imager/Spectral Radiometer (DISR) Aboard Huygens," pp. 109-138 in *Huygens Science, Payload and Mission*, SP-1177, European Space Agency, Noordwijk, The Netherlands, August 1997. Also see, M.G. Tomasko, L.R. Doose, P.H. Smith, C. Fellows, B. Rizk, C. See, M. Bushroe, E. McFarlane, E. Wegryn, E. Frans, R. Clark, M. Prout, and S. Clapp, "The Descent Imager/Spectral Radiometer (DISR) Instrument Aboard the Huygens Probe of Titan," reprinted from *Cassini/Huygens: A Mission to the Saturnian Systems, Proceedings from the International Society for Optical Engineering*, August 5-6, 1996, Denver, Colo., Vol. 2803, pp. 64-74.

2.4 INFORMATION THAT IMPROVES MISSION DESIGN

Every flight project is the culmination of an investment in research on the ground. Effective missions could not be designed without this research funded by R&DA programs. Earth-based telescopic observations have helped in the selection of landing sites on the Moon and Mars and entry sites for atmospheric probes (Example 1); atmospheric models were essential in designing aeroshells and parachutes for the Pioneer Venus, Galileo, and Cassini-Huygens programs; cosmic dust collected on aerogel in the stratosphere by U2 aircraft earned the proposed Stardust comet flyby credibility as a comet-sampling mission; ground-based, airborne, and rocket-borne solar astronomy evolved to become the sophisticated Solar and Heliospheric Observatory (SOHO) mission; and modeling, theory, and field investigations of the microwave brightness of the oceans led to operational instruments, such as the Special Sensor Microwave/Imager (SSM/I), for estimating ocean-surface winds (Example 2). Other examples are given in Box 2.4.

1. Earth-based observations of planets are used to plan new missions and provide the scientific context for data returned by probes. Mission planners for a planetary probe must anticipate atmospheric and, for a lander, surface conditions that the spacecraft will encounter. Scientists who interpret data returned from the spacecraft also want to know how representative the landing or entry site is of the entire planet and whether the environmental conditions are subject to change. Investigations that provide the pre-mission estimates of expected conditions are funded by R&DA programs.

Infrared images and spectra have been crucial for inferring temperatures, pressures, and compositions to relatively great depths in the atmospheres of the jovian planets (Jupiter, Saturn, Uranus, and Neptune) and for monitoring martian surface conditions. Primary sources of these data have been the Hubble Space Telescope; the R&DA-supported Infrared Telescope Facility (IRTF) at Mauna Kea, Hawaii (Plate 2.4); and the R&DA-supported Kuiper Airborne Observatory. The Mars Pathfinder (Mars rover) mission and the highly successful Galileo probe, which performed the first in situ measurements of Jupiter's atmosphere, were absolutely dependent on R&DA-funded, ground-based observations for their mission designs and for placing flight data in a larger scientific context.[16]

2. Satellite measurements of ocean-surface winds improve global weather forecasts. Remote sensing science and the technologies that permit retrieval of ocean-surface winds from active and passive microwave data are largely products of R&DA programs. Ocean-surface winds are operationally assimilated in the ocean-atmosphere energy- and moisture-exchange models that are coupled with atmospheric models to predict weather. Before satellite data became available, the surface-wind measurements for the models were from scattered ships of opportunity and sparse networks of weather buoys. R&DA-funded research established the relationship between microwave brightness and ocean roughness and contributed to our understanding of the relationship between ocean roughness and wind speed. The combination of these two relationships translates microwave brightness into an estimate of wind speed. The reliability of the combined relationship was demonstrated with brightness data from NASA's Scanning Multichannel Microwave Radiometer (SMMR) on NOAA's Nimbus 7 satellite.

With a successful heritage from SMMR, the Defense Meteorological Satellite Program began including the SSM/I on its satellites in 1987. These represent the first generation of operational, global,

---

[16]For further reading, see G. Orton et al., "Earth-Based Observations of the Galileo Probe Entry Site," *Science* 272:839-840, 1996.

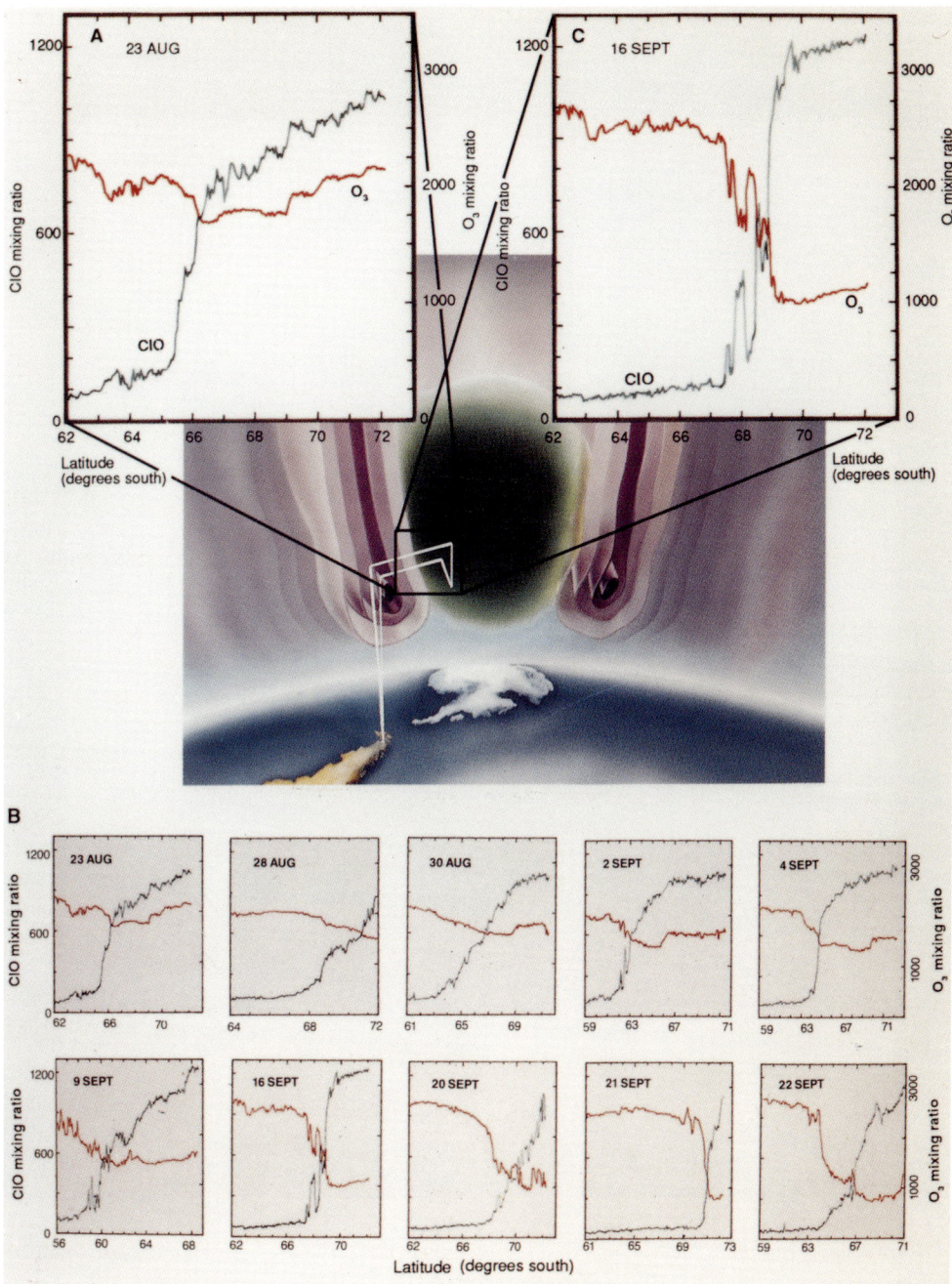

PLATE 2.1 Rendering of the containment provided by the circumpolar jet that isolates the region of highly enhanced ClO (shown in green) over the antarctic continent. Evolution of the anticorrelation between ClO and O_3 across the vortex transition is traced from (A) the initial condition observed on August 23, 1987, on the southbound leg of the flight, (B) the summary of the sequence over the 10-flight series, and (C) the imprint on O_3 resulting from 3 weeks of exposure to elevated levels of ClO. Data panels do not include the dive segment of the trajectory; ClO mixing ratios are in parts per trillion by volume, and O_3 mixing ratios are in parts per billion by volume. SOURCE: J.G. Anderson, D.W. Toohey, and W.H. Brune, "Free Radicals Within the Antarctic Vortex: The Role of CFCs in Antarctic Ozone Loss," *Science* 251:39-46, 1991.

PLATE 2.2 Hubble deep-field image. SOURCE: Courtesy of R.E. Williams, Space Telescope Science Institute.

PLATE 2.3 Muriel Ross (center) and Rei Cheng (left) use an immersive virtual environment workbench and special "Crystal Eyes" glasses to study a three-dimensional image of a skull. The skull was reconstructed from a computer tomography scan using special software developed in the NASA Ames Biocomputation Center. Interaction with images can be achieved using a pencil-like pointer (stylus) or special gloves as shown here.
SOURCE: Courtesy of M.D. Ross, NASA Ames Research Center.

PLATE 2.4 NASA Infrared Telescope Facility, Mauna Kea, Hawaii. SOURCE: Courtesy of Robert D. Joseph and John Rayner, NASA Infrared Telescope Facility Institute for Astronomy, University of Hawaii, Honolulu. Photograph by Richard Wainscoat, 1991.

> **Box 2.4**
> **Other Examples of Information That Improves Mission Design**
>
> - R&DA-funded research contributed to the first ground-based measurements of solar oscillations. The study of these oscillations is called helioseismology. R&DA projects also produced the theoretical models of the solar interior and solar dynamo that are used to interpret helioseismic data.
> - The discovery of solar oscillations inspired the design of many investigations of the Sun using SOHO and TRACE instruments.[1]
>
> ---
>
> [1] For further reading, see National Research Council, Space Studies Board, *Assessment of Programs in Solar and Space Physics*, National Academy Press, Washington, D.C., 1991.

ocean-surface-wind mappers. Global, near-daily image data from these satellites are routinely available through NOAA and the National Snow and Ice Data Center in Boulder, Colorado.

Initial attempts to use these surface-wind estimates in numerical weather-forecasting models were not successful because of inaccuracies in retrieved wind direction and in the relationship between surface winds and winds higher in the boundary layer. Further R&DA-funded research led to a more reliable wind-retrieval algorithm and to better modeling of surface and higher-level winds. Validation tests of the new retrieval algorithm showed rms (root-mean-square) accuracies better than 3 knots.[17] The new algorithm was used to generate a 9-year, global, ocean-surface wind data set that has become a significant resource for climate research.

More recent R&DA-funded research contributed to development of the scatterometer-based retrieval of ocean-surface winds that is used operationally by NOAA to estimate surface winds from ERS-1 scatterometer data. It also led to the development of the NASA Scatterometer (NSCAT).[18] Aircraft validation of NSCAT data has shown as little as a 1-knot rms error in estimates of wind speed. NSCAT was a key instrument on the failed Advanced Earth Observing Satellite (ADEOS) mission. A follow on scatterometer, Seawinds, will fly on NASA's Quickscat mission and a second SeaWinds instrument will fly on ADEOS II.[19]

---

[17] In this case, accuracy = $\left(\sum_n (x_n - x_a)^2\right)^{1/2}$, where x_n is the nth estimate of wind speed and x_a is the actual wind speed.

[18] For further reading, see R. Atlas, S.C. Bloom, R.N. Hoffman, E. Brin, J. Ardizzone, J. Terry, D. Bungato, and J.C. Jusem, "Geophysical Validation of NSCAT Winds Using Atmospheric Data and Analyses," *Journal of Geophysical Research*, March 1998; Robert Atlas, "Atmospheric Observations and Experiments to Assess Their Usefulness in Data Assimilation," *Journal of the Meteorological Society of Japan* 75 (1B):111-130, 1997; Robert Atlas, "Preliminary Evaluation of NASA Scatterometer Data and Its Application to Ocean Surface Analysis and Numerical Weather Prediction," reprinted from *Earth Observing Systems II, Proceedings of the International Society for Optical Engineering*, 28-29 July 1997, San Diego, California, Vol. 3117, pp. 90-97.

[19] For further reading on the theory of remote sensing of the ocean, see L.-L. Fu, W.T. Liu, and M.R. Abbott, "Satellite Remote Sensing of the Ocean," pp. 1193-1236 in *The Sea: Ocean Engineering Science*, Vol. 9, Wiley Interscience, New York, 1990; A. Stoffelen and D.L.T. Anderson, "Ambiguity Removal and Assimilation of Scatterometer Data," *Quarterly Journal of the Royal Meteorological Society* 123:491-518, 1997; and R. Atlas, "Atmospheric Observations and Experiments to Assess Their Usefulness in Data Assimilation," *Journal of the Meteorological Society of Japan* 75(1B):111-130, 1997.

2.5 INVESTMENTS THAT INCREASE THE PRODUCTIVITY OF FLIGHT PROJECTS

Flight projects generally support a specific set of activities that are limited in time and scope. R&DA programs maximize the scientific return from flight projects through preparatory research prior to the initiation of flight projects, supporting research during missions, post-mission analyses after flight projects end, and synthesis research using data from more than one flight. Example 1 shows laboratory measurements funded by R&DA programs being used to extract information about planetary atmospheres from routine spacecraft radio signals. Examples 2 and 3 illustrate R&DA-funded research continuing to extract significant scientific results from space data long after the missions ended.

1. Laboratory measurements of microwave absorption in simulated planetary atmospheres are used to recover information about planetary atmospheres from the signal strengths and frequency distortion of communications signals from spacecraft. Nearly all planetary spacecraft transmit science and operations data to Earth by radio. As a spacecraft that is transmitting to Earth passes behind the limb of a planet or moon that has an atmosphere, the radio signal transits increasingly deeper paths in the atmosphere until it is lost by absorption and scattering in the atmosphere or blocked by liquids and solids that lie below the atmosphere. During periods when the signal transits the atmosphere, spectral components of the signal may be absorbed by atmospheric gases or distorted by Doppler shifts in frequency when scattered by gas molecules moving coherently (wind) or randomly (temperature). If the dielectric properties of the candidate atmospheric gases are known, radioabsorption data can be interpreted as vertical profiles of temperatures and gas abundances in the extraterrestrial atmosphere.

The relevant dielectric data for each of the candidate atmospheric gases must be measured in a laboratory under realistic atmospheric conditions. R&DA funds have been used to develop a simulation chamber and then to build the library of measurements. This "radio science" approach to probing planetary atmospheres was used on the Voyager, Pioneer-Venus, Magellan, Galileo, and Mariner missions to derive profiles for the abundances of the atmospheric gases H_2SO_4, SO_2, SO_3, H_2O, NH_3, H_2S, PH_3, and CH_4. These profiles are used to constrain models of chemical and dynamical processes that govern the atmospheric systems.[20]

2. Synthesis of data from Sun-Earth connection Explorer missions yields key discoveries after these missions end. The International Sun-Earth Explorer (ISEE) mission consisted of three spacecraft: ISEE-1 and 2, launched into the same highly elliptic Earth orbit in October 1977, were designed to observe the dynamic near-Earth space; ISEE-3, launched in August 1978, was stationed in front of Earth to observe the solar wind destined to strike Earth's magnetic field. Although the prime mission of the ISEE spacecraft spanned 3 years—from 1977 to 1980—the ISEE-1 and 2 spacecraft were operated for 7 more years as extended missions before they reentered Earth's upper atmosphere. During these extended missions, data analysis funds were available for studying all of the data collected by ISEE. Many of these analysis projects focused on data from the first several years of flight, leaving much of the later data unanalyzed. After the missions ended, grants to analyze ISEE data could be sought from the R&DA program and such grants are still being funded.

The prime mission, 1977-1980, was not the time of greatest ISEE-related scientific activity (Figures 2.5 and 2.6). The number of publications from the magnetospheric data peaked in 1984 (with a second, lower peak in 1993) and has decreased slowly since then (Figure 2.5). Although the prime and extended

---

[20]For further reading, see G.F. Lindal, "The Atmosphere of Neptune: An Analysis of Radio Occultation Data Acquired with Voyager 2," *Astronomical Journal* 103:967-982, 1992; A.J. Kliore et al., "The Ionosphere of Europa from Galileo Radio Occultations," *Science* 277:355-358, 1997.

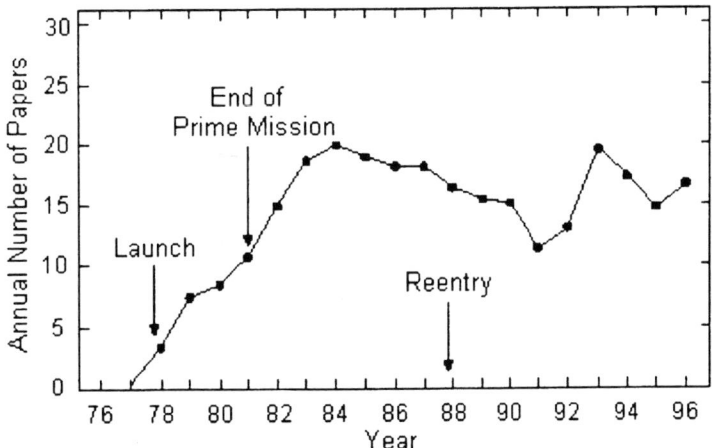

FIGURE 2.5 Papers in journals and books analyzing the International Sun-Earth Explorer magnetic field data. SOURCE: Courtesy of C.T. Russell, University of California, Los Angeles.

missions eventually spanned nearly an entire 11-year solar cycle, it was the extended mission that captured the disturbances at the height of the solar activity. In fact, some of the most significant results were obtained in the 1990s when coronal mass ejections were recognized as the origin of major "space weather" storms experienced in the vicinity of Earth. This discovery revolutionized the Sun-Earth connection approach to studying geomagnetic activity and its consequences and became the basis for National Space Weather Program forecasts about the effects of solar disturbances.

The ISEE record illustrates that mission-funded data analysis during an extended mission and R&DA-funded data analysis after the extended mission were essential in the continuing discovery of fundamental aspects of the behavior of the magnetospheric plasma.[21] These grants maintained the community's access to the data and brought into the program students, postdoctoral researchers, and guest investigators with new ideas. The ISEE program repeatedly demonstrated that it is not always possible to recognize the ultimate importance of data as they are being collected. Many of the important scientific results were considered controversial when first presented because they defied conventional wisdom.[22]

3. SEASAT continues to provide excellent examples of synthetic aperture radar images 20 years after the "failed" mission. SEASAT was launched into a polar Earth orbit on June 28, 1978. On October 10, 1978, the satellite suffered a massive short circuit in its electrical system and stopped functioning. The satellite carried four microwave sensors: (1) a radar altimeter, (2) a microwave scatterometer, (3) an L-band synthetic aperture radar (SAR), and the Scanning Multichannel Microwave Radiometer. SEASAT objectives were to collect data on sea-surface winds and temperatures, wave heights, internal waves, atmospheric water vapor, sea-ice features, and ocean topography.

---

[21]*International Sun-Earth Explorers Interim Bibliography,* National Space Science Data Center, Goddard Space Flight Center, Greenbelt, Md., 1987.

[22]For further reading, see P. Song, C.T. Russell, J.T. Gosling, M. Thomsen, and R.C. Elphic, "Observations of the Density Profile in the Magnetosheath Near the Stagnation Streamline," *Geophysical Research Letters* 17:2035-2038, 1990; P. Song, C.T. Russell, and M.F. Thomsen, "Slow Mode Transition in the Frontside Magnetosheath," *Journal of Geophysical Research* 97:8295-8305, 1992.

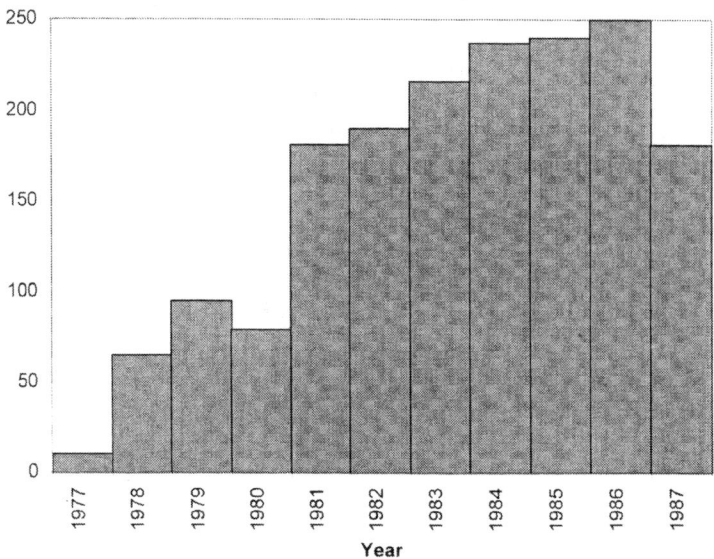

FIGURE 2.6 Number of International Sun-Earth Explorer published papers, 1977-1987. SOURCE: Adapted from data provided by K. Olgilvie, NASA Goddard Space Flight Center.

Despite the brevity of SEASAT's 100-day lifetime, SEASAT data have been employed by various R&DA programs over the past 20 years with impressive results. The topographic data were processed to make the world's first gravity map of seafloor features based on satellite altimetry. Although the map had low resolution, it did show trenches, midoceanic ridges, fracture zones, and other tectonic features, and it served the community for a decade until measurements from ESA's ERS-1 were made public in 1995 and the Navy declassified Geosat data.

Although the sensors were designed for measurements over water, the SEASAT radar altimeter acquired more than 600,000 useful altimeter range measurements over the continental ice sheets of Greenland and Antarctica.[23] These data now serve as an important baseline for temporal comparisons with data from current sensors such as NSCAT. There has been nearly complete SEASAT SAR coverage of the continental United States, and these data continue to provide a base map for detecting geologic structure and changes in land cover (Figure 2.7).

SEASAT data were used widely to familiarize the planetary science community with the strengths and weaknesses of photographic interpretation of geologic features with SAR data in anticipation of the Magellan images of Venus. SEASAT data allowed planetary geologists to become familiar with radar image distortion and the competing effects on radar backscatter of roughness, dielectric contrast, incidence angle, and subsurface scattering. Even after the Magellan data were in hand, comparisons with SEASAT data continued to be used in analyses.

---

[23]For further reading, see H.J. Zwally, A.C. Brenner, J.A. Major, R.A. Binschadler, and J.G. Marsh, "Growth of Greenland Ice Sheet: Measurement," *Science* 246:1587-1589, 1989; T.V. Martin, H.J. Zwally, A.C. Brenner, and R.A. Bindschadler, "Analysis and Tracking of Continental Ice Sheet Radar Altimeter Waveforms," *Journal of Geophysical Research* 88 (C3):1608-1616, 1983; and C.H. Davis, C.A. Cluever, and B.J. Haines, "Elevation Change of the Southern Greenland Ice Sheet," *Science* 279:2086-2088, 1998.

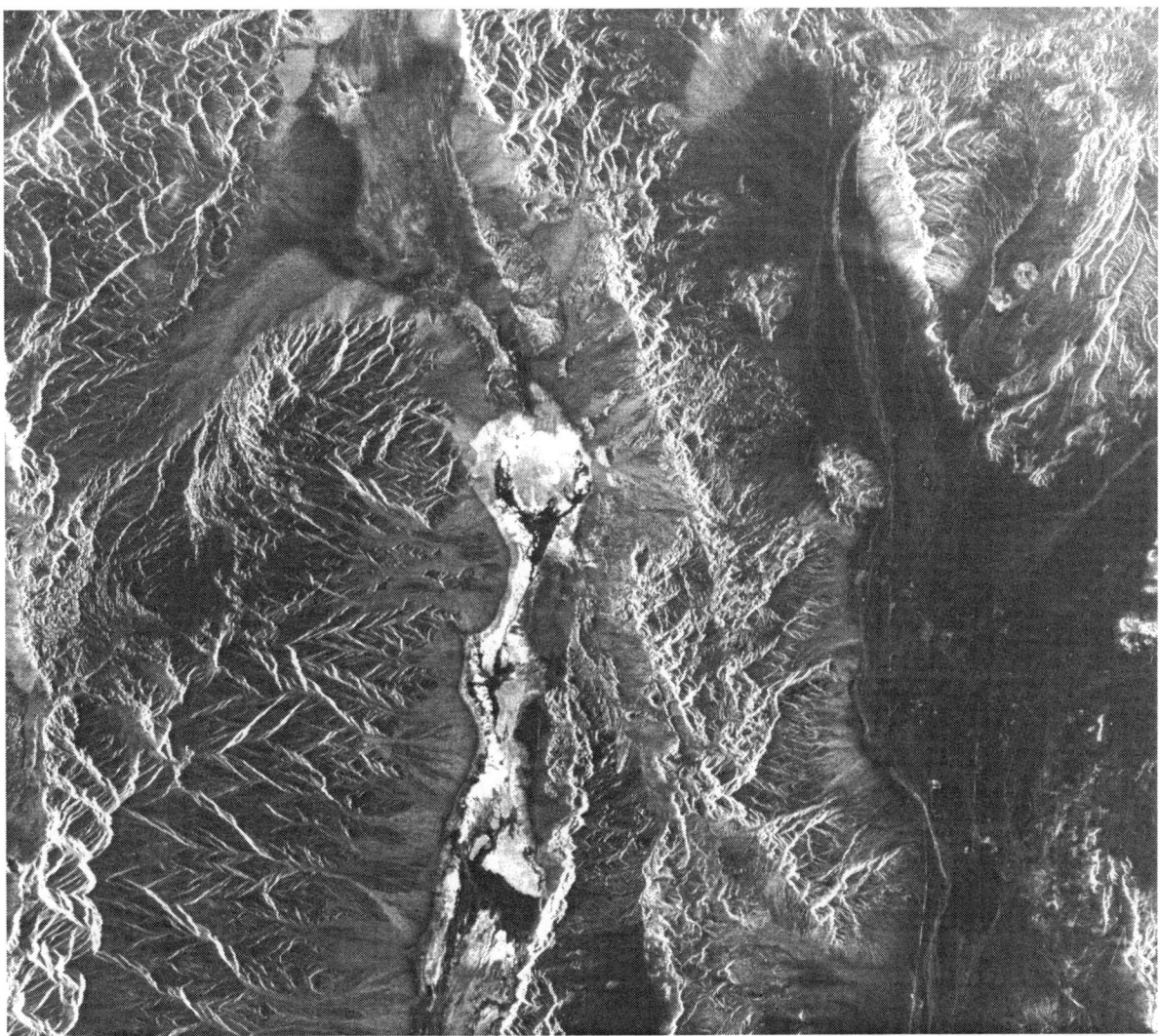

FIGURE 2.7 Image of the Death Valley, California, region acquired on September 14, 1978, by the SEASAT synthetic aperture radar (SAR). West is at the top of the scene, which is dominated by the mountains of the Panamint Range. Alluvial fans flow eastward from the mountains into the valley. The valley floor is characterized by very rough salt deposits that appear bright, and smooth playas appearing dark. The valley is bordered on the east by the Amargosa Range. The flat area east of the Amargosa Range is the Amargosa Desert. The Amargosa River and the low mountain ranges of western Nevada are in the bottom portion of the image. SOURCE: Courtesy of the Jet Propulsion Laboratory Radar Data Center.

Interferometric synthetic aperture radar, a new technology for performing high-resolution topographic mapping, has stimulated recent interest in the SEASAT SAR data. Retrospective use of these data provides some of the best examples available on which scientists can test results of interferometric processing algorithms. The continuing exploitation of SEASAT's data set for planetary, oceanographic, and land-surface research is a testament to the importance of R&DA funds for continued data analysis in select cases where the data are unique.

2.6 RESEARCH THAT COMPLEMENTS THE WORK OF OTHER FEDERAL AGENCIES

The interests of NASA and other federal agencies often appear to overlap in the areas of space science, but because of the different focus and expertise of various agencies, the efforts are in fact complementary. Complementarity is further ensured by external advisory committees that span the interests of both agencies. In instances where the divisions of effort are not clear, agencies often coordinate their efforts with joint research announcements. Recent examples include programs jointly funded by NASA and NSF for studies of the Shoemaker-Levy 9 comet and its impact on Jupiter, and studies of the ALH84001 meteorite from NSF's antarctic collection and of great importance to NASA's Mars exploration program. Other examples of complementarity and cooperation can be found in joint efforts with the National Institutes of Health (NIH) in space medicine, with NOAA in terrestrial and space weather, with the Environmental Protection Agency (EPA) in ecosystem studies, and with the Department of Defense in weather and oceanic studies. Yet another example of collaborative efforts is described briefly in Box 2.5.

1. Cooperation between the National Institute of Standards and Technology (NIST) and NASA may lead to improved manufacturing processes for metals. Dendrites are tiny crystalline structures that form inside molten metals and metal alloys when they freeze. Nearly all industrially important metals solidify dendritically from the molten state. An understanding of how these dendrites form can lead to improvement in alloy strength and ductility and/or to a reduction in production costs.

The morphology and growth kinetics of dendrites are sensitive to gravity-driven convective flows in melts. The R&DA program for microgravity research has been supporting comprehensive experimental investigations of dendritic growth in the absence of gravity. For example, an investigation that began at the Naval Research Laboratory evolved into shuttle-based experiments by a team of researchers from Rensselaer Polytechnic Institute to test the generally accepted, approximate Ivantsov model's prediction that the shapes of dendrite tips were paraboloids of revolution. Results from the space-based experiments are in conflict with the prediction and invite a reexamination of the underlying physical processes.[24] The academic investigators are working with theoreticians at NIST to develop a dendrite-growth model that is consistent with the new data.[25]

2. The U.S. Department of Agriculture's (USDA's) space-based crop-monitoring programs began with NASA collaboration. After launch in the early 1970s of the first of the series of multispectral scanners on Sun-synchronous satellites that was to become the Landsat program, NASA and the USDA began collaborative research that used data from these sensors to estimate crop yield and crop condition. Through their joint efforts, the technology evolved that enabled USDA's current operational programs to, for example, monitor rice, cotton, and soybeans in the Mississippi Delta, corn in the Midwest, and

---

[24]M. Glicksman, R.J. Schaefer, and J.D. Ayers, "Dendritic Growth—A Test of Theory," *Met. Trans.* 7A:1747-1759, 1976; M. Glicksman, M.B. Koss, and E.A. Winsa, "The Isothermal Dendritic Growth Experiment: A Fundamental Test of Theory in Advanced Materials," *Trans. Mat. Res. Soc., Japan* 16A:611-619, 1994; and M. Glicksman, M.B. Koss, L.T. Bushnell, J.C. LaCombe, and E.A. Winsa, "Space Flight Data from Isothermal Dendritic Growth Experiment," *Adv. Space Res.* 7 (27):181-184, 1995.

[25]For further reading, also see S.R. Coriell, G.B. McFadden, R.F. Boisvert, M.E. Glicksman, and Q.T. Fang, "Coupled Convective Instabilities at Crystal-Melt Interfaces," *J. Crystal Growth* 66:514-524, 1984; R.N. Smith, M.E. Glicksman, and M.B. Koss, "Effects of Buoyancy on the Growth of Dendritic Crystals," *Annual Review of Heat Transfer* VII, 1995; M.E. Glicksman, M.B. Koss, and E.A. Winsa, "Dendritic Growth Velocities in Microgravity," *Phys. Rev. Lett.* 73:573, 1994; and R.F. Sekerka, S.R. Coriell, and G.B. McFadden, "The Effect of Container Size on Dendritic Growth in Microgravity," *J. Crystal Growth* 171:303-306, 1977.

> **Box 2.5**
> **Another Example of Research That Complements**
> **the Work of Other Federal Agencies**
>
> NASA joins the NSF, NOAA, and DOE in the U.S. Global Change Research Program through participation in projects such as the Global Energy and Water Cycle Experiment (GEWEX).[1] To date, most of NASA's participation has been through R&DA-funded field experiments and the development of the tools of remote sensing science. Eventually, these R&DA investments will enable regional and global investigations of Earth's climate systems.
>
> ---
> [1] For additional information on GEWEX, see the GEWEX home page at <http://www.cais.com/gewex.html>.

wheat on the Prairie. The USDA's Foreign Agricultural Service, whose concern is the world productivity of economically important crops, has become the largest user of Landsat data. Its Landsat-based analyses are used by Congress, agribusiness, the State Department, the Agency for International Development, and the United Nations.

Collaboration between the USDA and NASA was initiated with several programs: first the Large Area Crop Inventory Experiment (LACIE) and then the Agriculture and Resources Inventory Surveys Through Aerospace Remote Sensing (AgRISTARS). The technologies for land-cover classification developed under LACIE were the foundation for the operational USDA program. Crop-yield and crop-condition estimation was developed under AgRISTARS.[26]

2.7 SCIENCE-DRIVEN ADVENTURE THAT STIMULATES INTEREST IN MATH, SCIENCE, OR ENGINEERING EDUCATION

As Example 1 on Mars meteorites in Section 2.2 suggests, small R&DA grants can attract broad (even worldwide) public attention, much like the attention given major flight projects. Normally, these grants represent a quiet activity that nevertheless serves as the primary link between NASA and academic research in most universities most of the time. Opportunities for participation in flight projects are more rare and highly competitive.

R&DA grants provide the support base for NASA-funded academic research, the primary means for training graduate students in space-related disciplines, and opportunities for exposing undergraduates to aerospace science and engineering. R&DA-funded suborbital programs (i.e., balloon, aircraft, and sounding rocket investigations) offer students and young investigators training opportunities that prepare them for greater responsibilities on later spaceflight projects. The relatively low cost and rapid turnaround of suborbital flights are better suited to doctoral research than are most spaceflight projects in which predoctoral students are unlikely to be given significant responsibilities and project duration often exceeds student Ph.D. tenures. Furthermore, the potentially high frequency of suborbital flights permits the young investigator to build a record of successful experiments.

---

[26] For further reading, see R.B. MacDonald, F.G. Hall, and R.B. Erb, *The Use of Landsat Data in a Large Area Crop Inventory Experiment (LACIE)*, National Aeronautics and Space Administration, Houston, Tex., 1975; *Agriculture and Resources Inventory Surveys Through Aerospace Remote Sensing: AgRISTARS*, prepared by AgRISTARS Program Support Staff, National Aeronautics and Space Administration, Houston, Tex., 1981.

In addition to the educational function of R&DA grants to academia, NASA sponsors a series of outreach opportunities that use the dramatic nature of space-related research to stimulate the interest of students in kindergarten through college in science and math. The frequent links to academic research are illustrated in three examples.

1. The NASA Space Grant College and Fellowship Program exposes K-16 students to space science and technology. The NASA Space Grant College and Fellowship Program, established in 1989, administers awards to state-based consortia in three areas: research, education, and public service.[27] The NASA contribution is approximately doubled through matching funds from consortia members. The funds are used to build stronger educational elements where research in space science and technology exists and to introduce space science and technology where they have not been established.

For example, Space Grant funds for the South Dakota Space Grant Consortium have been used largely for educational outreach for K-12 programs.[28] Highlights of the Space Grant Consortium's educational components include the following:

- University student design, development, and launch of scientific payloads on high-altitude balloons and analysis of data from these flights;
- South Dakota Space Day, which attracted more than 5,000 K-12 students from across the state;
- A summer aerospace workshop for K-12 teachers;
- Student scholarships in science, which have expanded the educational opportunities for underrepresented groups including women, American Indians, and African Americans; and
- Summer fellowship awards, which have allowed faculty to focus on space-related science each summer.

This has been accomplished with a NASA annual commitment of less than $200,000.

2. NASA's Minority University Research and Education Division (MURED) works with historically black colleges and universities and other minority universities to expand contributions to NASA's science and technology base. NASA outreach to these minority institutions increases diversity among science and engineering students and encourages early involvement of minority faculty in NASA-related research. These MURED grants are competitively selected, peer-reviewed R&DA-type awards.

The first MURED program created university research centers (URCs) at minority institutions in 1991.[29] Each URC, although established by NASA Headquarters, is managed by a designated NASA field center. The URCs are intended to increase the competitive capabilities among minority colleges and universities in aerospace research, expand the nation's pool of aerospace research and development institutions, increase minority institutions' faculty and student involvement in research, and increase the number of students from underrepresented groups with advanced degrees in NASA-related disciplines.[30]

University Research Centers at minority institutions include the following:

- Alabama A&M University (Center for Hydrology, Soil Climatology and Remote Sensing),

---

[27] For a brief description of this program, see "A Guide to NASA Education Programs" available on NASA's Education Resources home page at <http://www.hq.nasa.gov/office/codef/education/html.index>.

[28] For additional information, see the South Dakota Space Grant Consortium home page at <http://www.sdsmt.edu/space/space.html>.

[29] For additional information, see NASA's Minority University Research and Education Division's University Research Centers at Minority Institution's home page at <http://mured.alliedtech.com/pub/www/awards/urc>.

[30] For more information, see <http://www.hq.nasa.gov/office/codee/mured.html> on the World Wide Web.

- Clark Atlanta University (High Performance Polymers and Composites Center),
- Fisk University (Center for Photonic Materials and Devices),
- Florida A&M University (Center for Nonlinear and Nonequilibrium Aeroscience),
- Hampton University (Center for Optical Physics),
- Howard University (Center for the Study of Terrestrial and Extraterrestrial Atmospheres),
- Morehouse School of Medicine (Space Medicine and Life Sciences Research Center),
- North Carolina A&T State University (Center for Aerospace Research),
- Prairie View A&M University (Center for Applied Radiation Research),
- Tennessee State University (Center for Automated Space Science),
- Tuskegee University (Center for Food and Environmental Systems for Human Exploration of Space),
- University of New Mexico (Center for Autonomous Control Engineering),
- University of Puerto Rico at Mayagüez (Tropical Center for Earth and Space Studies), and
- University of Texas at El Paso (Center for Earth and Environmental Studies).

URCs received $4.1 million in FY 1996 and $2.1 million in FY 1997.

3. *The Global Learning and Observations to Benefit the Environment (GLOBE) program brings the experiences of environmental research to the K-12 classroom.* NASA, NOAA, and NSF have teamed up to create an international program, GLOBE, that will encourage environmental scientists and teachers to share the experience of discovery with K-12 students. GLOBE projects are intended to increase the following:

- Students' scientific understanding of Earth,
- Their achievements in science and mathematics, and
- Their environmental awareness.

GLOBE currently enlists students in 1,500 schools to make environmental observations following established research protocols. These students report their data via the Internet/World Wide Web to the GLOBE Student Data Archive where the data are publicly available. An additional 2,000 schools have committed to follow these same protocols and begin similar observations.

GLOBE offers training and online information for teachers who wish to participate in the program. It also maintains an online forum where students from all over the world can discuss environmental observations and issues.[31]

2.8 SUMMARY COMMENTS

These notable examples of the contributions of the R&DA programs help explain why the programs are important to the nation, to science, to the advancement of technology, and to education. Recognizing the difficulty in communicating the value of R&DA to those both inside and outside NASA, the task group chose what it considered to be successful examples of R&DA. It believes that this is one of the first summaries of the accomplishments of NASA's R&DA programs.

---

[31]For further reading, see "GLOBE: A Worldwide Environmental Science and Education Partnership," *Journal of Science Education and Technology*, Vol. 7, No. 1, 1998; information is also available on the GLOBE Web site at <http://www.globe.gov>.

3

The Role of the Research and Data Analysis Programs

Chapter 1 affirms the central role given science by NASA's charter, by top-level reviews of its mission,[1] and by policy statements from the White House.[2] The task group also introduces the theme of this report in Chapter 1—that vigor and quality in NASA's science enterprise require healthy R&DA programs as well as numerous flight projects; this theme is illustrated in Chapter 2. This chapter discusses critical science questions as one of the foundations of R&DA, differences and commonalities in R&DA across the agency, the importance of linking R&DA to NASA's strategic plans, and how R&DA is responding to a changing environment—for example, the impact that policies to streamline and shorten missions can have on both the expectations and the resource requirements for R&DA activities.

3.1 UNDERSTANDING THE BASIS OF R&DA

NASA's science is organized under three enterprises: (1) the *space science* enterprise managed by the Office of Space Science (OSS); (2) the *Earth science* enterprise managed by the Office of Earth Science (OES); and (3) the *human exploration and development of space* enterprise managed by the Office of Life and Microgravity Science and Applications (OLMSA) and the Office of Spaceflight (OSF). (See Appendix B for a diagram of NASA's organizational structure.) Each office has developed or is developing a strategic plan that includes its priorities in science. These priorities can be cast as critical science questions and used to guide the agency's science programs. Summaries of current sets of critical questions appear in the following sections.

---

[1] National Commission on Space, *Pioneering the Space Frontier*, Bantam Books, May 1986; *Report of the Advisory Committee on the Future of the U.S. Space Program*, U.S. Government Printing Office, Washington, D.C., December 1990; Space and Earth Science Advisory Committee, NASA Advisory Council, *The Crisis in Space and Earth Sciences*, November 1986.

[2] National Science and Technology Council, "National Space Policy," The White House, September 1996.

3.1.1 Critical Science Questions for the Space Science Enterprise

The space science enterprise encompasses the traditional scientific disciplines of astronomy and astrophysics, space and solar physics, and planetary science. Through investigations in these areas, NASA seeks to answer fundamental scientific questions, some of which are as old as the human race (Box 3.1). These questions have significance well beyond the scientific community; they lie at the heart of humanity's attempt to understand its place in the universe. OSS also investigates fundamental physical and biological laws using space environments as natural laboratories that cannot be duplicated on Earth.

The tools of the space science enterprise are extraordinarily broad and varied. They include deep-space probes; Earth-orbiting astrophysical observatories; aircraft-, balloon-, rocket-, and ground-based observatories; and laboratory and theoretical investigations. These tools, used in concert, have facilitated understanding of complex questions about the universe and the solar system.

3.1.2 Critical Science Questions for the Earth Science Enterprise

The Earth science enterprise consists of one program—Earth science—whose primary objective is to understand the interactions among Earth's land, oceans, and atmosphere that influence weather, climate, Earth's ecosystems, agriculture, and hazards to populations. This cross-cutting approach to Earth studies has come to be known as Earth system science. Critical science questions from NASA's earth science strategic plan are listed in Box 3.2.

A sense of urgency emerged in the Earth sciences as investigators began to identify the strong linkages between phenomena as diverse as the depletion of stratospheric ozone and the terrestrial use of chlorofluorocarbons (CFCs); severe storms and the El Niño-Southern Oscillation; and the growth of infrared-active gases in the atmosphere and potential climate changes that will affect human health, economic decisions, fishery yields, and agricultural productivity. Progress in Earth system science is sensitively tied to the research strategy selected to attack these problems.[3]

3.1.3 Critical Science Questions for the Human Exploration and Development of Space Enterprise

Unlike the space and Earth science enterprises, the human exploration and development of space (HEDS) enterprise encompasses much that lies beyond concerns usually attributable to science. These range from issues as grand as the expansion of human life beyond Earth to issues as practical as the commercialization of access to space. OLMSA manages the HEDS research program, which spans biology, medicine, materials science, fluid and combustion physics, and biotechnology.

Life Sciences Research

Critical questions within the life sciences largely concern the effects of gravity (or the absence of gravity) on plant, animal, and human physiological systems. For example, a primary question is, How do humans adapt to the space environment and readapt on return to Earth?

---

[3]National Research Council, Board on Sustainable Development, *Overview of Global Environmental Change: Research Pathways for the Next Decade*, National Academy Press, Washington, D.C., 1998.

> **Box 3.1**
> **NASA Space Science Enterprise Strategic Plan**
>
> **Fundamental Questions[1]**
>
> - How did the universe begin, and what is its ultimate fate?
> - How do galaxies, stars, and planetary systems form and evolve?
> - What physical processes take place in extreme environments such as black holes?
> - How and where did life begin?
> - How is the evolution of life linked to planetary evolution and to cosmic phenomena?
> - How and why does the Sun vary, and how do Earth and the other planets respond?
> - How might humans inhabit other worlds?
>
> **NRC Research Strategy Questions**
>
> **Astronomy and Astrophysics[2]**
>
> - To what extent are the origin and evolution of life a consequence of the evolution of the solar system?
> - How did life arise on Earth, and are we humans unique?
> - How old is the universe?
> - What is the geometry and mass of the universe?
>
> **Space and Solar Physics[3]**
>
> - What are the mechanisms of solar variability?
> - Can we predict solar variability?
> - What is the physics behind the solar wind and the heliosphere?
> - What are the structure and dynamics of magnetospheres, and what is their coupling to adjacent regions?
> - What are the dynamics of the middle and upper atmospheres, and what is their coupling to regions above and below?
> - What are the plasma processes that accelerate very energetic particles and control their propagation?
>
> **Planetary Sciences[4]**
>
> - How did the solar system originate?
> - How have its constituents evolved?
> - How, in general, do planets work?
>
> ---
>
> [1]*The Space Science Enterprise Strategic Plan: Origins, Evolution, and Destiny of the Cosmos and Life*, National Aeronautics and Space Administration, Washington, D.C., November 1997, p. 4.
> [2]National Research Council, Space Studies Board, *A New Science Strategy for Space Astronomy and Astrophysics*, National Academy Press, Washington, D.C., 1997.
> [3]National Research Council, Space Studies Board, *A Science Strategy for Space Physics*, National Academy Press, Washington, D.C., 1995.
> [4]National Research Council, Space Studies Board, *An Integrated Strategy for the Planetary Sciences: 1995-2010*, National Academy Press, Washington, D.C., 1994.

> **Box 3.2**
> **Questions from NASA's Earth Science Strategic Plan[1]**
>
> - Can climate variation be predicted a season or year in advance?
> - Can long-term climate variations be detected and drivers identified?
> - What are the impacts of climate change on marine ecosystems?
> - How do terrestrial ecosystems respond to land cover change?
> - How do sudden solid Earth changes affect the land surface?
>
> ---
> [1] *Mission to Planet Earth Strategic Enterprise Plan 1996-2002*, National Aeronautics and Space Administration, Washington, D.C., March 1996.

This question is related to physiological changes facing astronauts during extended space travel. Even relatively short spaceflights produce physiological alterations that include bone demineralization, cardiovascular deconditioning, muscular atrophy, immune dysfunction, and postflight vestibular disruptions. Although apparent in many astronauts during short spaceflights, these changes have generally remained below the level of a clinical problem. With increasingly longer exposures to a microgravity environment, such changes may become clinically pronounced. There are also issues of limiting exposure to ionizing radiation and the maintenance of mental health during long missions. Critical science questions for the life sciences are listed in Box 3.3.

Microgravity Research

Microgravity research focuses on understanding physical and chemical processes in the reduced-gravity environment. Not only are these processes central to the successful exploration and development of space, but the removal of gravitational constraints may also open new avenues for progress in science and technology. The specific systems and processes of interest span fluid dynamics and transport phenomena, materials science, combustion science, biotechnology, and low-temperature physics. Critical science questions are listed in Box 3.4.

3.2 UNDERSTANDING THE ROLES OF R&DA

The role of R&DA programs—their relationship to other components of a science program—is obscured by the differing definitions and content of R&DA across NASA program offices (Box 3.5). Although these differences are explained, in part, by variations in the character of the science among science disciplines, some of the differences are clearly arbitrary. Similarly, the budget categories for R&DA-type activities are fragmented; funding can reside within the R&A program, the advanced technology development (ATD) program, the MO&DA program, or the suborbital program.[4] The term "R&DA" used in the text of this report and in the data presented in Chapter 4 includes these elements (ATD, MO&DA, R&A, and suborbital programs). For many observers inside and outside the agency

---
[4] The title "Supporting Research and Technology" was used by the Office of Space Science during the 1980s to encompass most of the programs included in R&DA. Use of the older title is avoided here to emphasize the complementary role of R&DA to flight programs rather than any "supporting" role.

> **Box 3.3**
> **NRC Research Strategy Questions[1]**
>
> **Physiological and Psychological Effects of Spaceflight**
>
> - What are the physiological mechanisms that lead to bone and muscle deterioration during long-term spaceflight, and how can mechanistic insights be used to develop effective countermeasures?
> - What are the bases for the adaptive compensatory mechanisms in the vestibular and sensorimotor systems that operate both on the ground and in space?
> - What are the magnitude, time course, and mechanisms of cardiovascular adjustments for long-duration exposure to microgravity?
> - What are the mechanisms underlying inadequate total peripheral resistance observed during postflight orthostatic stress?
> - What are the carcinogenic risks following irradiation by protons and high-charge Z and high-energy particles?
> - Will exposure to heavy ions at the level that would occur during long-duration deep-space missions pose a risk to the integrity and function of the central nervous system?
> - Can the combination of radiation and stress on the immune system that occurs in space produce additive or synergistic effects?
> - What role does the host response to stressors during spaceflight play in alterations in host defenses?
> - What are the neurobiological and psychosocial mechanisms underlying the effects of physical and psychosocial environmental stressors during spaceflight?
>
> **Graviperception and Gravitropism in Plants**
>
> - Which cells actually perceive gravity in a plant, and what are the intracellular mechanisms by which the direction of the gravity vector is perceived?
> - What is the nature of the cellular asymmetry that is set up in a plant cell that perceives the direction of the gravity vector?
> - What are the nature and mechanism(s) of the translocation of the signal(s) in plants that pass from the site of perception to the site of reaction?
> - What is (are) the mechanism(s) by which gravitropic signals in plants cause unequal rates of cell elongation, and what are the possible effects of gravity on the sensitivity of these cells to the signals?
>
> **Animal Graviperception, Reproduction, and Development**
>
> - What role does gravity play in normal development of the gravity-sensing vestibular system of animals?
> - How does microgravity influence the development and maintenance of neural space maps in the brain?
> - Are there developmental processes in vertebrates that are critically dependent on gravity?
>
> ---
> [1]This list is a sampling of types of studies recommended in National Research Council, Space Studies Board, *A Strategy for Space Biology and Medicine in the New Century*, National Academy Press, Washington, D.C., 1998.

> **Box 3.4**
> **Overarching Research Questions in the Microgravity Sciences[1]**
>
> • What are the microscopic and macroscopic effects of gravity on the physical and chemical processes associated with natural phenomena and human activities?
> • How does gravity affect the common processes found in natural or industrial activities?
> • What other physical mechanisms become dominant as the effective gravitational acceleration is reduced to a very low level?
> • Can gravity be used as an adjustable parameter to perform controlled basic scientific and engineering experiments and to learn how to improve current technological processes?
>
> ---
>
> [1] The overarching research questions noted for microgravity research are based on previous discipline research priorities developed by the NRC: National Research Council, Space Studies Board, *Toward a Microgravity Research Strategy*, National Academy Press, Washington, D.C., 1992; National Research Council, Space Studies Board, *Microgravity Research Opportunities for the 1990s*, National Academy Press, Washington, D.C., 1995.

who have not been students of the R&DA budgets, it has been difficult to correlate the impact of R&DA budget and policy changes with NASA science. There is no simple relationship between the agency's budget components and the goals for R&A outlined for each NASA science program office. The task group tried to bring the various programmatic elements together with their budgets as a first step toward linking budget trends to changing priorities.

Despite the disjointed nature of R&DA, the task group and the Space Studies Board isolated a set of components they consider to be common, fundamental elements of R&DA programs. These components are (1) theoretical investigations; (2) new instrument development; (3) exploratory or supporting ground-based and suborbital research; (4) interpretation of data from individual or multiple space missions; (5) management of data; (6) support of U.S. investigators who participate in international missions; and (7) education, outreach, and public information. Their role, in part, is not unlike the sequential discipline of hypothesis, test, and synthesis referred to as the scientific method.

1. *Theoretical investigations.* Theory plays a unifying role in the quest to understand nature by identifying important observational or experimental questions and providing a coherent framework for observations that may at first appear unrelated. Theory includes the development of numerical models and computer simulations that facilitate the broadest feasible applications of data from individual observations and the development of predictive capabilities. R&DA strategies support theoretical investigations that complement experimental programs.

2. *New instrument development.* Many of the instruments that have enabled research in space have been developed under R&DA programs by small groups of innovative researchers in NASA centers, universities, and industry. Even instruments that fly on major missions often have design and proof-of-concept heritages that were funded by R&DA programs. R&DA funding for instrument development assumes even greater importance for smaller, faster, cheaper and technology-driven missions. This applies equally to robotic spacecraft and piloted platforms such as the International Space Station (ISS).

> **Box 3.5**
> **The Diverse Character of NASA's R&A Activities (FY 1999)[1]**
>
> The goals of the *space science* R&A program are to (1) enhance the value of current space missions by carrying out supporting ground-based observations and laboratory experiments; (2) conduct the basic research necessary to understand observed phenomena and develop theories to explain observed phenomena and predict new ones; and (3) continue the analysis and evaluation of data from laboratories, airborne observatories, balloons, rocket experiments, and spacecraft data archives. In addition to supporting basic and experimental astrophysics, space physics, and solar system exploration research for future flight missions, the program also develops and promotes scientific and technological expertise in the U.S. scientific community.
>
> The goals of the *Earth science* applied R&DA program are to advance our understanding of the global climate environment and the vulnerability of the environment to human and natural forces of change, and to provide numerical models and other tools necessary for understanding global climate change. The applied R&DA program is divided into two components: (1) Earth science program science and (2) Earth science operations, data retrieval, and storage. The activities under program science include research and analysis, Earth Observing System (EOS) science, airborne science and applications, the purchase and management of scientific data, commercial remote sensing, and the uncrewed aerial vehicle (UAV) science program. Operations, data retrieval, and storage include several independent activities responsible for the operation of currently functioning spacecraft and flight instruments, high-performance computing and communications, and the provision of computing infrastructure.
>
> Commencing with the FY 1999 congressional budget submission, the OLMSA budget structure has been realigned to reflect the reorganization of programmatic activities into five programs and three functions. Therefore, the *life sciences* R&A program is now divided into programs for advanced human support technology (AHST), biomedical research and countermeasures (BR&C), and gravitational biology and ecology (GB&E):
>
> • The goals of the AHST program are to (1) demonstrate and validate full self-sufficiency in air and food recycling technology for use in space vehicles; (2) demonstrate and validate integrated, fully autonomous environmental monitoring and control systems; and (3) validate and incorporate human factors engineering technology and protocols to ensure the maintenance of

3. *Exploratory or supporting ground-based and suborbital research.* Many disciplines require extensive supporting ground-based research to make effective use of space-based opportunities. Examples include ground-based and airborne observations of planets or galaxies, ground-based validation of satellite observations, airborne sampling of the atmosphere, and drop-tower experiments of microgravity effects.

4. *Interpretation of data from individual or multiple space missions.* Major missions were once expected to fund extensive data analysis. With current policies that favor many much smaller missions making more limited observations and with the ascendancy of the ethic that publicly funded space data should rapidly be made public, data synthesis from one or more space missions to validate or refute hypotheses is increasingly the responsibility of R&DA programs.

5. *Management of data.* Careful stewardship of data and information is central to maximizing near- and long-term payoffs from R&DA-sponsored programs. Data management includes creating cata-

> **Box 3.5 Continued**
>
> high ground and flight crew skills during long-duration missions. The AHST program makes NASA technologies available to the private sector for Earth applications.
>
> - The BR&C program develops understanding of the underlying mechanisms of the effects of spaceflight on humans. Its applied research activities also develop countermeasures to prevent the undesirable effects of spaceflight on humans. The program includes several areas of research: space physiology, environmental health, radiation health, operational medical research, and behavior and performance. The overriding goal of these activities is to enable the human exploration and development of space by minimizing risks and optimizing crew safety and performance.
> - Finally, GB&E focuses on research designed to improve our understanding of the role of gravity in biological processes from the cell to global ecosystems. The emphasis in this program is on advancing fundamental knowledge in the biological sciences, but the research supported often also contributes to the other goals of the HEDS enterprise. The program solicits research in molecular, cellular, developmental, organismal, population, and comparative biology that seeks an understanding of basic mechanisms underlying the effects of gravity on these systems. NASA continues to value ground-based research leading to flight experiments that can confirm or refute the fidelity of ground-based models and hypotheses.
>
> The *microgravity science* R&A program seeks to understand basic physical phenomena and processes; quantify effects and overcome limitations imposed by gravity on the observation and evaluation of selected phenomena and processes; develop technologies related to research requirements; and expand, centralize, and disseminate the research database as widely as possible to the U.S. research and technology community. The primary goals of the microgravity research program are to advance fundamental scientific knowledge of physical, chemical, and biological processes; to enhance the quality of life on Earth by conducting scientific experiments in the low-gravity environment of space; and to mature the research of a large number of laboratory scientists into coherent groups of flight experiments.
>
> ---
>
> [1]"Science, Aeronautics, and Technology FY 1999 Estimates Budget Summary," National Aeronautics and Space Administration, Washington, D.C.

logues for data, conducting inventories of the data, and archiving the data. The success of data management programs can be measured by the ease with and rate at which the research community can access the data.

6. *Support of U.S. investigators who participate in international missions.* NASA has a long record of productive international collaborations with other nations and groups of nations.[5] The European Space Agency's (ESA's) Giotto mission, the Japanese Yohkoh mission, the Soviet Vega mission, the German Roentgen satellite, and the ESA Infrared Space Observatory are only a few of the many successful collaborative programs. R&DA programs have helped bridge research gaps in U.S. programs by supporting U.S. investigators' participation in international missions.

---

[5]National Research Council and European Science Foundation, *U.S.-European Collaboration in Space Science*, National Academy Press, Washington, D.C., 1998.

7. *Education, outreach, and public information.* The drama and adventure of science in space attract students to science and mathematics. Whether or not these students continue with space-related studies, the benefit to the nation of having a populace that is scientifically literate is unarguable. NASA's outreach programs and its R&DA programs are naturally complementary in that R&DA programs are often the source of the information that fuels outreach activities, and students who are influenced by outreach programs often find themselves working on R&DA projects first as undergraduates, then as research assistants in graduate school, and later as postdoctoral researchers.[6]

3.3 POSING A STRATEGY FOR R&DA PROGRAMS

Although the contributions of R&DA activities to science and to the nation may be demonstrable, as illustrated in Chapter 2, the role and importance of R&DA to NASA and other agencies have been difficult to describe. Some agency and government officials have viewed the programs as "entitlements" for scientists. Moreover, decision makers report that they lack an appreciable understanding of R&DA, its direction, and its performance or output.

Without question, the task of linking measurable and quantifiable outputs to R&DA activities is difficult and echoes attempts to find "metrics" for government-funded research and development. However, the link between R&DA and the agency and enterprise strategic plans is one means by which to assess R&DA's performance and direction. NASA's science strategic plans are all built on a sequence of key elements—goals, objectives, implementation approaches,[7] priorities, performance measures—that define the strategy for the program. The fundamental underpinnings of the strategy are critical science questions that the science program seeks to answer; these questions constitute the basis for the goals of the program. Increasingly, agency strategic plans are also focused on the applications of space sciences to practical needs. These applications, if they are to be effective, must be anchored in a strong science base.

R&DA provides an especially important and powerful means of ensuring that a strategy's goals, objectives, and critical science questions are rooted in good science and that they evolve to reflect scientific progress. It influences all elements of a strategy. For instance, R&DA serves as the platform from which to develop and improve implementation approaches (e.g., via development of new technologies) and through which the results of the program are extracted (e.g., via data analyses) to create meaningful results, performance measures, and progress in achieving initial goals and objectives. When programs are inadequately anchored by critical science questions, we find projects whose primary objectives are the collection of data; programs that continue without significant discoveries or advances; and results that can be forgotten without penalty. The more the R&DA activities are integrated into the strategy and managed with a view to enhancing the implementation and evolution of the strategy, the stronger is the overall program.

The effectiveness of R&DA's contribution to a strategy relies on the balance and health of R&DA programs as a whole. R&DA can be evaluated by examining the extent to which it contributes to progress in implementing the strategic plan. For example, one can ask, What is the impact of R&DA on refining critical questions, defining new and better approaches and technologies, answering critical

---

[6]The NASA Office of Space Science Web page provides details on education and outreach as well as links to several related space entities. See <http://www.hq.nasa.gov/office/oss/education/index.htm>.

[7]By "implementation approaches" is meant the portfolio of tools in implementing NASA's strategy, such as large and small flight missions; dedicated missions and flights of opportunity; spaceflights and suborbital programs; ground-based research; and principal investigator, consortia, and team-oriented projects.

questions (data analysis), and placing answers within a scientific context (e.g., theory)? Other measures of R&DA contributions to NASA and the scientific community might entail exploring whether R&DA is sustaining an adequately sized research community, generating new ideas, identifying forward-looking technologies, providing infrastructure to permit widespread use of flight data, and providing training for new researchers and engineers.

The task group recognizes that serendipitous discoveries often depend on investigators having flexibility within their grants to explore anomalous observations or occasionally follow their curiosity. The task group believes that this is the nature of a research grant as opposed to a contract and does not wish the distinction to be lost. Within this context, however, it is important that R&DA projects be linked to critical science questions and that the linkages be stronger as project costs increase.

3.4 RESPONDING TO THE CHANGING ENVIRONMENT

NASA's efforts to streamline and simplify missions so as to increase the flight rate also provide increased opportunities for investigators to access spaceflight data. Shorter-duration missions also allow effective responses to exciting new science discoveries, the rapid infusion of new technologies, and opportunities for involvement by young scientists. Examples of NASA programs employing this approach include the Earth System Science Pathfinder (ESSP) program for applications and the Small Explorer (SMEX) and Discovery programs for solar system exploration. The task group supports these approaches and notes the Explorer program as an exemplary case (Box 3.6).

The agency's move to smaller and shorter-duration flight programs has also introduced new demands on R&DA programs. For instance, much of the science and some of the technologies previously developed under lengthy flight projects will now be funded out of the research base. Similarly, those who submitted proposals to the ESSP program were directed to request only the funds necessary to collect and validate flight data; more general analyses of the data were to be provided through projects

---

Box 3.6
Smaller, Faster, Cheaper and the Explorer Program

The Explorer program played a crucial role in the early development of NASA's programs in astronomy, astrophysics, and solar and space physics. Focused on low-cost missions that could respond to scientific opportunities in a timely manner, the Explorer program was well suited to disciplines dominated by new discoveries. As the pressures to exploit discoveries with more sophisticated capabilities exceeded the resources available to support new starts for major missions, a number of large Explorers were approved and initiated. However, the heavy fiscal demands of these Delta-class Explorers greatly reduced the opportunity to mount at least one new Explorer mission per year, resulting in a backlog of innovative ideas and a dearth of flight opportunities. Returning the Explorer program to its original focus, by restricting it to small- and medium-class missions (SMEX and MIDEX, respectively) that are strictly limited in cost, has restored the opportunity for two to three missions per year and has resulted in a number of exciting new scientific investigations. This revitalization of the Explorer program is one of the best examples of the benefits of the smaller, faster, cheaper philosophy.

funded by R&DA programs.[8] Previously, a significant fraction of data analyses were funded by flight programs. Other emerging stresses on the R&DA program include the trend away from long flight project development periods, which allowed the development of project-unique technologies. Some flight projects in the past even carried definition-phase funds to develop needed technologies. Now shorter flight projects may be expected to be ready for launch within 3 years of project approval. In such a compressed schedule there is no time to develop new technologies. Thus, mission-specific technologies are now expected to be funded out of the research base.

The role of R&DA in supporting elements of the nation's scientific infrastructure is often obscured by NASA's image as a "mission agency." R&DA programs are primary sources of support for scientists outside NASA whose work benefits from access to space or use of aerospace technologies, and they are the dominant sources of support for NASA's in-house scientists. Not only are these programs of discovery, but they also educate each new generation of researchers in space-related sciences, engineering, and project management skills, and they stimulate interest in science and mathematics for many others. For space-related sciences, NASA's responsibility is equivalent to that of the National Science Foundation or the National Institutes of Health for disciplines that rely primarily on ground- rather than space-based observations.

In the face of flat or decreasing NASA science budgets, it is important to examine whether the agency's R&DA program can adequately meet the combination of traditional and new roles over the full range of scientific disciplines and research institutions.

---

[8]NASA Announcement of Opportunity, Earth System Science Pathfinder Missions (ESSP), AO-96-MTPE-01, July 19, 1996, p. 13.

4

Budget Trends for the Research and Data Analysis Programs

The objective of this chapter is to examine budget data for trends in the 1990s in funding of R&DA programs; in allocation of R&DA funds among NASA laboratories, academia, and industry; and in size and duration of grants. The task was laborious, and the risk of inaccuracy was high because R&DA programs are dispersed through many budget categories, definitions within budget categories vary among NASA's three science program offices, and these definitions have changed from year to year.

Faced with these difficulties, the task group limited its analyses to three sources of published data: (1) NASA's budget books for FY 1991 through 1998, (2) NSF's sectoral overview of basic research funding, and (3) NASA's grants and contracts books. The budget books offer the most accurate numbers but provide surprisingly little insight about who did the work or, in fact, about how funds were actually spent. The NSF overview offers the task group's only insight into who did the work, but the numbers are aggregated at too large a level to reflect only R&DA allocations. The grants and contracts books reveal much about how funds were used by universities and, to some extent, by industry, but they provide no information about their use by NASA centers.

These three sources offer a limited window on NASA's R&DA activities, but because they represent different cuts at a complex problem, they are not perfectly consistent among themselves. That is, readers will find small differences among what might otherwise appear to be equivalent budget categories in the three data sources. Even within a single data source, certain categories were modified by the task group to achieve greater year-to-year consistency. For example, Hubble Space Telescope (HST) operations have been separated from HST MO&DA because the task group is aware that some MO&DA funds were used to build second-generation HST instruments.

4.1 OVERALL NASA FUNDING TRENDS FOR R&DA: FY 1991-1998[1]

NASA budget books for FY 1991 through FY 1998 were used to examine gross budget trends. Budget data, along with relevant language from budget justification documents and experience with NASA programs, allowed the task group to track R&DA-related activities through changes in accounting definitions.

The task group's definition of the NASA "research base" pulls together six categories of data extracted from the NASA budget: (1) traditional R&A line items; (2) data collected on an accumulation of relevant ground-based activities, referred to as "other science support," that have only recently begun to appear in the NASA budget; (3) suborbital programs; (4) MO&DA; (5) supporting infrastructure, a category devised to incorporate engineering, operations, and other support for science infrastructure;[2] and (6) academic programs, as specifically listed in NASA budget documents. Box 4.1 defines these six budget categories. The budget trends for these and other components of NASA science-related activities during the 1990s are given in Table 4.1.

It is difficult to get a sense of the historical trends for the DA funding of MO&DA because NASA has not separated DA from MO&DA. The task group notes that the rapid increases in MO&DA early in the decade were driven by the operational needs of large space science missions such as the HST and the Gamma Ray Observatory. The summary here does not reflect the full impact of these MO&DA increases since the HST operations and servicing line have been moved to the flight hardware budget summary. (See Appendix A, Table A.2) It is important to note that NASA is implementing a major restructuring of these activities.[3]

The task group makes the following observations:

• Total R&DA funding (adjusted for inflation) grew 44 percent between FY 1991 and FY 1998, but the increase was due mainly to growth in the Earth Observing System Data and Information System (EOSDIS) (with an increase of $157 million in FY 1995 dollars over the 8-year period) and to the transfer of technology development funding from the Office of Space Access and Technology (OSAT) to OSS ($201 million in FY 1995 dollars). R&DA funding in areas other than EOSDIS and the transfer of technology funds from OSAT showed a net increase of about 7 percent in inflation-adjusted terms.

• Traditional R&A declined by $98 million (in FY 1995 dollars), or nearly 22 percent.

• All science-related, ground-based activities account for about 11 percent of the overall NASA budget in FY 1998, up from about 7 percent in FY 1991.

• Funding for the suborbital program increased 19 percent in constant dollars. The increase was related to the transition from the Kuiper Airborne Observatory to the new Stratospheric Observatory for Infrared Astronomy (SOFIA).

• Academic outreach programs increased from a small base of about $61 million for distributed activities in FY 1991 to $113 million in FY 1998.

---

[1]The task group is grateful for the assistance of Mr. Malcolm Peterson, comptroller, National Aeronautics and Space Administration, and his staff in providing some of the data on NASA budgets. Any errors are attributable to the task group and not to NASA.

[2]The detailed historical tables in Appendix A provide information on the science discipline breakdown that has been the traditional program structure of the NASA budget. Specific R&A budget line items can be found in earlier budgets for physics and astronomy (P&A, which includes astrophysics and space and solar physics), for planetary exploration, for life sciences and microgravity research, and for the Earth sciences.

[3]"Full Costing in NASA," Office of Chief Financial Officer, National Aeronautics and Space Administration, February 1996.

Box 4.1
Components of NASA Ground-based Research

Research and Analysis

The traditional R&A NASA budget line items differ across NASA program offices. For example, in the Office of Space Science (OSS), R&A supports designs for future missions, sensor and instrument development, and ground-based observations and experiments, among other activities. In life sciences, the R&A program supports applied and basic research in biomedicine, biology, environmental science, and related technologies, including ground-based research, as well as support facilities and technologies, among other items. In microgravity science, R&A includes ground-based experiments; experiments selected for flight; and research in biotechnology, combustion science, fluid physics, materials science, and low-temperature physics. The Earth science applied R&DA program supports the Office of Earth Science (OES) science and operations, data retrieval and storage, science for Earth Observing System (EOS) programs, airborne science, and applications, among other activities.[1]

Other Science Support

Other science support includes mission studies and technology development, a new category of the OSS budget; EOS science; and mission science teams and guest investigator programs in the OES budget.

Suborbital Programs

The category "suborbital programs" includes funding for sounding rockets, high-altitude balloon flights, and the operation of NASA's fleet of space- and Earth-science research aircraft.

MO&DA

MO&DA includes funding for the operations of data-collecting hardware, analysis of data, satellite operations during core missions, and continuation of data analysis after the core mission. Preflight preparations and preliminary data analysis are also supported under this category. It does not include HST operations and servicing funds, which were grouped by the task group as major flight project funding because of the large element of instrument development funding included in this line item. (See Appendix A, Table A.2.)

Supporting Infrastructure[2]

Included in the category "supporting infrastructure" is funding for a variety of activities that support NASA science programs but are not directly focused on the performance of space research (e.g., engineering, operations, facilities support).

Academic Programs

Academic programs include NASA training grants (all academic levels) and minority research and education programs.

---

[1] NASA FY 1995 budget submission to Congress.

[2] The supporting infrastructure line in Table 4.1 is not part of the NASA formal budget structure, but this grouping has been used to collect various items in the NASA budget that are essential support items for the research program but differ somewhat from the R&DA activities that are the central focus of this study. For a more detailed look at the content of the infrastructure category and the general framework of historical budget statistics prepared for this study, see Appendix A, which summarizes NASA congressional budget submissions over the entire period from FY 1981 through FY 1998.

TABLE 4.1 NASA Research and Analysis Budgets in Context, FY 1991-1998

NASA Science-related Programs and Activities	FY 1991	FY 1992	FY 1993	FY 1994	FY 1995	FY 1996	FY 1997	FY 1998[a]	Percent Change, FY 1991-1998
In Millions of Constant FY 1995 Dollars									
R&A (traditional definition)	453	426	431	438	430	417	388	355	−21.5
Office of Space Science	184	158	182	183	184	153	160	122	−33.4
Office of Life and Microgravity Science and Applications	78	86	74	75	81	83	86	79	2.3
Office of Earth Science	191	183	175	179	165	181	142	154	−19.8
Other science support	0	0	0	0	64	46	81	83	N/A
EOS Data and Information System (EOSDIS)	40	84	137	193	221	242	224	197	394.1
Suborbital programs	83	87	90	97	93	113	76	99	19.3
MO&DA (adjusted)	372	413	479	429	379	401	403	371	0.0
Supporting infrastructure	5	54	78	52	67	81	103	44	787.1
Science-related technology programs	0	0	0	0	0	0	192	201	N/A
Academic programs	61	72	97	88	106	107	115	113	84.6
Total ground-based activities	1,014	1,134	1,312	1,297	1,359	1,408	1,581	1,464	44.4
As Percentage of NASA Budget									
R&A (traditional definition)	2.9	2.8	2.9	2.9	3.1	3.1	3.0	2.8	
Other science support	0.0	0.0	0.0	0.0	0.5	0.3	0.6	0.6	
EOS Data and Information System (EOSDIS)	0.3	0.5	0.9	1.3	1.6	1.8	1.7	1.5	
Suborbital programs	0.5	0.6	0.6	0.7	0.7	0.8	0.6	0.8	
MO&DA (adjusted)	2.4	2.7	3.2	2.9	2.7	3.0	3.1	2.9	
Supporting infrastructure	0.0	0.3	0.5	0.3	0.5	0.6	0.8	0.3	
Science-related technology programs	0.0	0.0	0.0	0.0	0.0	0.0	1.5	1.6	
Academic programs	0.4	0.5	0.6	0.6	0.8	0.8	0.9	0.7	
Total ground-based activities	6.6	7.4	8.7	8.7	9.7	10.4	12.1	11.4	
As Percentage of Total NASA Science-related Activities									
R&A (traditional definition)	16.5	14.2	13.9	11.9	10.9	10.8	11.0	10.0	
Other science support	0.0	0.0	0.0	0.0	1.6	1.2	2.3	2.3	
EOS Data and Information System (EOSDIS)	1.4	2.8	4.4	5.3	5.6	6.3	6.3	5.6	
Suborbital programs	3.0	2.9	2.9	2.6	2.4	2.9	2.1	2.8	
MO&DA (adjusted)	13.5	13.8	15.5	11.7	9.6	10.4	11.4	10.5	
Supporting infrastructure	0.2	1.8	2.5	1.4	1.7	2.1	2.9	1.2	
Science-related technology programs	0.0	0.0	0.0	0.0	0.0	0.0	5.4	5.7	
Academic programs	2.2	2.4	3.2	2.4	2.7	2.8	3.3	3.2	
Total ground-based activities	36.8	37.9	42.4	35.3	34.4	36.5	44.7	41.2	
Reference Data									
NASA total agency budget (million current dollars)	13,868	14,333	14,322	14,549	13,996	13,884	13,709	13,638	−2.7

TABLE 4.1 *Continued*

NASA Science-related Programs and Activities	FY 1991	FY 1992	FY 1993	FY 1994	FY 1995	FY 1996	FY 1997	FY 1998[a]	Percent Change, FY 1991-1998
Reference Data continued									
NASA total agency budget (million constant FY 1995 dollars)	15,358	15,428	15,028	14,922	13,996	13,571	13,111	12,800	–18.2
Total science-related activities (million current dollars)	2,486	2,781	2,947	3,583	3,950	3,951	3,695	3,781	51.7
Total science-related activities (million constant FY 1995 dollars)	2,753	2,993	3,092	3,675	3,950	3,862	3,534	3,548	27.6

NOTE: The data presented in Table 4.1 are based on appropriated budgets or budgets proposed for the congressional appropriations process. The numbers refer to "budget authority" rather than "outlays." That is, they represent funding levels approved for spending, but they are not necessarily equal to the actual expenditures in a given year for two reasons. First, there is usually a time lag between appropriations approvals and the actual outlay of funds. Second, there are usually small differences (typically a few percent) between the total appropriated level and the funds actually issued as grants and contracts in a particular account due to funds set aside for the congressionally mandated Small Business Innovative Research program and for some NASA institutional costs.

Several adjustments have been made to the data summarized in Table 4.1. In keeping with the objective of accounting for all R&DA-type funding, broadly defined, amounts for the Office of Space Science are included to accommodate the new category in NASA's recently restructured budget that is designated "mission studies and technology development." In a similar manner, the two categories in NASA's budget designated "EOS science" and "mission science teams and guest investigators" have been added to the Office of Earth Science. All three items are included in the category designated "other science support." In addition, the definition of the OSS MO&DA account has been narrowed by subtracting the "HST operations and servicing" component and moving these dollars to the summary of science-related flight projects. This latter adjustment was made because much of the funding in the HST operations and servicing budget is related to the development of advanced flight instrumentation for the Hubble Space Telescope.

Items may not add to totals because of rounding.
[a]Estimate.

Figure 4.1 illustrates trends in the NASA budget for ground-based research. The budget trends for R&A components of the science program offices are depicted in Figure 4.2 and show widely divergent patterns with some significant year-to-year variations. From FY 1991 to FY 1998 the OSS R&A budget decreased by 33 percent, the OES budget decreased by 20 percent, and the OLMSA budget decreased by 2 percent.

Following are some additional observations:

• From FY 1991 to FY 1998, R&A programs remained nearly constant as a proportionate share of the total NASA budget, hovering around 3 percent.

• The proportions for R&A as a share of total NASA science-related funding (the latter concept includes the total budgets of the three science program offices plus the academic programs) declined more than 6 percentage points, from 16.5 percent for FY 1991 to 10 percent projected for FY 1998—about a 35 percent share reduction in 8 years.

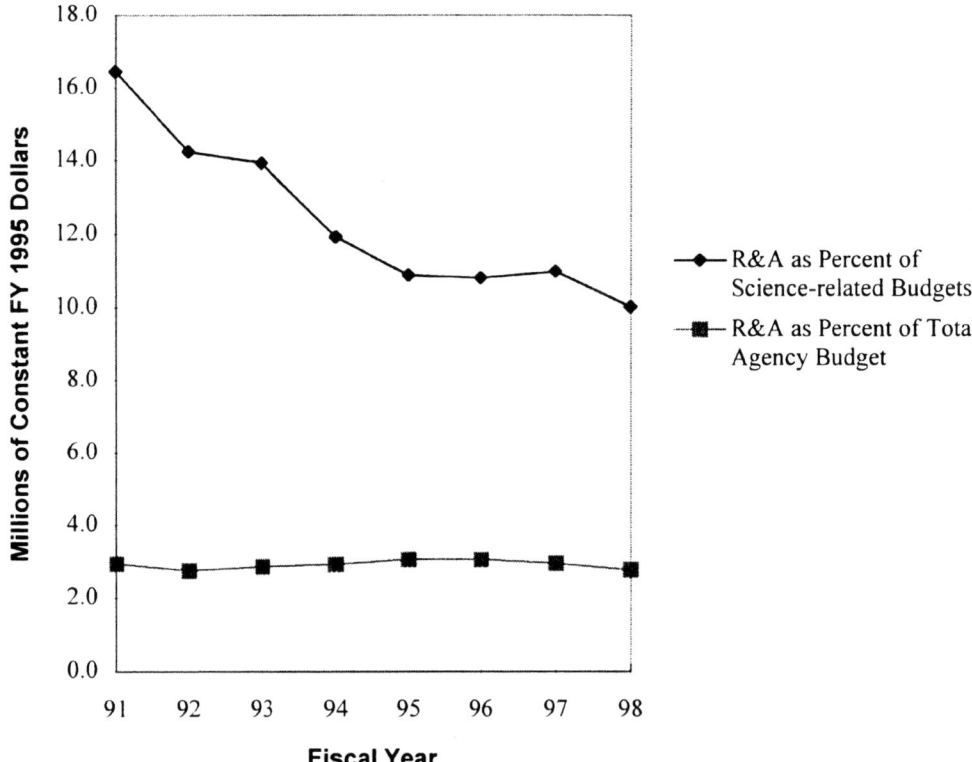

FIGURE 4.1 Trends in NASA budgets for ground-based research. Data for FY 1998 are for appropriations.

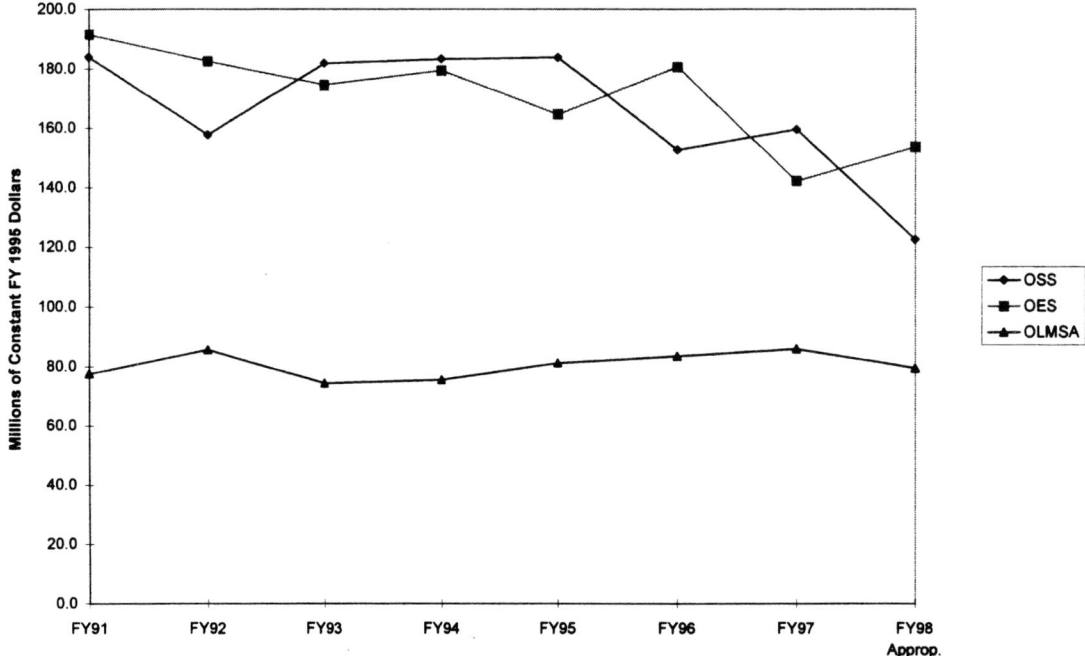

FIGURE 4.2 Budget trends in R&A components (traditional definition) for NASA's three science program offices.

4.2 DISTRIBUTION BY SECTOR OF NASA FUNDING FOR BASIC RESEARCH: FY 1991-1997

Each year, NASA submits estimates to the NSF about its research and development expenditures and where the work was performed. Expenditures are classified as basic research, applied research, or development. These data become part of NSF's annual report on federal support for research and development—the only published data of this type.

NASA's research-performing institutions are the federally funded research and development centers (FFRDCs), nonprofit institutions, NASA field centers, universities and colleges, and private industry. NASA's only FFRDC is the Jet Propulsion Laboratory (JPL) in Pasadena, California, which is the lead center for planetary programs, operates the deep-space network, and undertakes extensive technology development. Research funding for these five types of institution begins with the federal budget appropriation to NASA. The agency distributes the funds as procurements (contracts, grants, and cooperative agreements) to both internal and external scientists. Internal NASA research is currently funded both by institutional budgets for civil service salaries and facilities and by internal grants for research expenses associated with, for example, on-site contractors, hardware, or field work.

Table 4.2 provides a top-level look at NASA's allocations for basic research by performing sector. The distributions accounted for nearly $2 billion in research obligations for 1996 and included more than R&DA expenditures. For example, they included flight program and hardware procurements.[4] These data are the only currently published comprehensive estimates, as far as the task group is aware, of NASA's intramural (in-house) research program.

The task group makes the following observations:

• Colleges and universities constitute about one-quarter of the overall NASA basic research program. Basic research funding to academic institutions as a fraction of total NASA basic research rose from 21 percent in 1991 to 26 percent in 1997.

• NASA intramural and JPL funding constituted about 40 percent of the NASA research program in 1997, down from 45 percent in 1991.

• Although private industry constitutes about 30 percent of the research program, this may be misleading. Industry is less involved in independent research now than, for example, during the Apollo era. Some of industry's participation is likely to be contractor support to NASA field centers and to JPL. This would increase the NASA intramural and JPL share.

4.3 UNIVERSITY GRANTS AND CONTRACTS BY TYPE OF ACTIVITY: FY 1986-1995

Data in this section concern only the academic sector and were developed by the task group from extensive listings of contracts and grants awarded to universities.[5] Raw data are published annually by NASA Headquarters in a ledger format.[6] Because electronically scanning these ledgers to obtain data

---

[4] These data are statistical estimates prepared for NSF by various NASA entities as part of the annual budget process. The data do not have the same fidelity or basis in accounting records that is found in data contained in formal budget justification documents. These data, however, are useful in providing a broad perspective on how the agency spends its resources and particularly how the totals break down among the major research-performing institutions.

[5] NASA has published other, more accurate data about research performers in the academic sector. In searching for data about relative funding levels, statistics on average grant sizes, and other potential metrics for assessing research effectiveness and productivity, the task group found these only for science investigators at universities and colleges.

[6] NASA Headquarters publishes a large compendium frequently referred to as the NASA Green Book. These data were compiled into a database of statistics covering four fiscal years (FY 1986, FY 1989, FY 1992, and FY 1995).

TABLE 4.2 Distribution of NASA Basic Research Support by Performing Sector, FY 1991-1997

Sector	FY1991	FY1992	FY1993	FY1994	FY1995	FY1996	FY1997
In Millions of Constant FY 1995 Dollars							
Intramural	593	592	572	504	495	492	450
Industrial firms	709	572	545	725	636	653	587
Universities and colleges	478	483	453	432	480	470	459
Nonprofit organizations	67	72	68	62	60	59	54
FFRDCs—universities	424	400	332	266	301	288	246
FFRDCs—nonprofit	2	2	1	2	1	1	1
Total NASA	2,273	2,122	1,971	1,991	1,973	1,964	1,796
As Percentage of Total							
Intramural	26	28	29	25	25	25	25
Industrial firms	31	27	28	36	32	33	33
Universities and colleges	21	23	23	22	24	24	26
Nonprofit organizations	3	3	3	3	3	3	3
FFRDCs—universities	19	19	17	13	15	15	14
FFRDCs—nonprofit	0	0	0	0	0	0	0
Total NASA	100	100	100	100	100	100	100
Deflator (FY 1995 = 100)	74.8	83.3	92.9	97.5	100.0	102.1	104.7

NOTE: Latest actual data are for FY 1995; updated data are from the National Science Foundations's "advance tables" for volume 45, available on the NSF/SRS Web site. Details may not add to totals because of rounding.
SOURCE: National Science Foundation Federal Funds Reports (NSF97-302).

that could be manipulated was exceedingly labor intensive, the task group chose to limit its analyses to 3-year time increments from 1986 through 1995.[7]

R&DA funds are the primary means for the science community to participate in NASA programs. Although NASA keeps exhaustive records of individual procurement contracts and grants to universities, there has been no aggregation of these allocations by discipline, program office, or type of research activity. The task group believed that it was important to summarize, for the first time, the *uses of research funds* within the academic sector.

The task group developed the category of NASA "net space research" funding to the universities.[8] Net space research means funding for *research* from an R&DA source, as opposed to technology development, instrument development, and academic training that may be funded by other accounts. The method for classifying accounts in this way is detailed in Appendix A. Assignments of grants to disciplines or programs were based on the task group's familiarity with the research or the category of research that was funded by a particular program officer.

---

[7] Presenting some of the data for FY 1986 through FY 1995 and some for FY 1991 through FY 1998 may appear confusing and the data may be difficult to correlate; thus different intervals were chosen for the "award data" to get a historical sense of the budget trend.

[8] Appendix A describes and explains how net space research data were developed. The basic procedure entailed filtering out thousands of award records for each year to separate the largest awards, which were then hand-coded to reflect the types of activities carried out under each award. Smaller awards (generally less than $300,000 in any of the years examined) were then allocated on the basis of field-of-science codes that had previously been assigned by NASA technical and procurement personnel involved in awarding the specific contracts and grants.

TABLE 4.3 NASA Procurement Awards to Colleges and Universities, FY 1986-1995

Category	Funding (obligations in million constant FY 1995 dollars)				Composition (% of total)			
	FY 1986	FY 1989	FY 1992	FY 1995	FY 1986	FY 1989	FY 1992	FY 1995
Research Contracts and Grants								
OLMSA disciplines	22	25	34	45	6.1	4.5	4.8	5.4
OSS disciplines	85	122	151	136	23.6	22.2	21.1	16.4
OES disciplines	67	69	103	119	18.4	12.5	14.4	14.4
Subtotal—net space research[a]	174	216	289	299	48.1	39.2	40.2	36.3
Other Space Science Activities					0.0	0.0	0.0	0.0
Instrument design and development	39	68	86	119	10.8	12.3	11.9	14.4
Spacecraft design and development	16	23	31	48	4.5	4.2	4.3	5.8
Operation of science facilities	3	16	25	30	0.7	2.9	3.5	3.6
Operation of support facilities	12	20	8	9	3.4	3.6	1.1	1.1
Centers of excellence	2	1	3	5	0.4	0.1	0.5	0.6
Subtotal—other space science	72	127	153	210	20.0	23.1	21.3	25.5
Space Science-related Programs (sum of two categories above)	246	343	442	510	68.1	62.4	61.5	61.8
Other NASA Activities								
Training grants	11	16	31	43	2.9	2.8	4.2	5.2
National space grant colleges	0	0	14	15	0.0	0.0	1.9	1.8
Other education programs	7	10	12	30	1.8	1.8	1.7	3.7
Centers of excellence (nonscience)	5	30	46	41	1.4	5.4	6.4	5.0
Technology and technology transfer	78	133	140	151	21.7	24.1	19.5	18.3
Not Distributed	15	18	34	36	4.1	3.3	4.7	4.3
Subtotal—other NASA activities	116	207	277	316	31.9	37.4	38.4	38.3
TOTAL	362	550	719	826	100.0	100.0	100.0	100.0

NOTE: Details may not add to totals because of rounding.
[a]See Appendix A.

Table 4.3 lists NASA awards to colleges and universities. The first category indicates contracts and grants awarded by the three NASA science program offices: OLMSA, OSS, and OES. The second shows NASA funding for space-science-related activities other than bench-level research grants.[9] The most important of these are hardware design and development activities. The third category shows NASA awards that are not directly related to the performance of space research, including awards for various academic programs and activities associated with technology development and transfer. The largest of these are associated with aeronautical research and engineering programs.

---

[9]Universities also provide research support for NASA, including, for example, operation of tracking and data facilities or management of the National Scientific Balloon Facility.

NASA's definition of training grants, listed in the third category in Table 4.3, includes three segments of the NASA fellowship program: (1) the Graduate Student Research Program (GSRP) for students working on a dissertation project, (2) Global Change Research Fellowships for students concentrating in the area of global change, and (3) Minority Graduate Student Fellowships for minority students completing graduate work in space-related research. Students participating in the NASA fellowship program often work under the mentorship of an R&DA-funded professor.

The task group makes the following observations:

• There was a significant increase in net space research in the late 1980s corresponding, in part, to resumption of launch activity after the *Challenger* accident.

• During this same period, the percentage of net space research declined from 48 percent of total awards to universities in FY 1986 to 36 percent in FY 1995.

• Funding for nonscience centers of excellence and NASA technology development and transfer programs increased over the decade.

• Other science-related activities maintained approximately the same proportion of funding over the decade.

4.4 UNIVERSITY GRANTS AND CONTRACTS: AWARD SIZES AND DURATIONS

Its investment in developing a database of NASA procurement statistics enabled the task group to compute the average size and duration of awards to universities. Table 4.4 summarizes these data. New awards are also listed separately to show the effects on overall average award duration of the large number of NASA awards that are continued over long periods of time (in some cases a decade or longer). The median duration of new awards is 13 months and of continuing awards is 35 months. The median duration of all NASA grants was 35 months in FY 1995 compared to the historical median of 24 to 25 months in FY 1986, FY 1989, and FY 1992.

The task group makes the following observations:

• The number of annual awards by NASA to universities (both general awards and "net space research" awards) has expanded greatly since FY 1986, but most of this expansion seems to have occurred before FY 1992. The task group's data do not indicate the extent to which this growth reflects an increase in the number of investigators being funded versus an increase in the number of multiple grants being awarded to individual investigators.

• The use of simple mean values tends to be highly misleading in terms of what is actually happening to contract and grant sizes from the point of view of the typical NASA researcher. For example, while the average (mean) size of net space research awards (Table 4.4) to university recipients increased (in constant FY 1995 dollars) from $107,000 in FY 1986 to $113,000 in FY 1995, the most frequent award size (i.e., modal value) was $67,000 in FY 1986 and only $50,000 in FY 1995. The large difference between the modal value of all NASA university awards and the modal value of awards classified as net space research is accounted for by the several hundred training grants issued annually in NASA's graduate student fellowship programs.

• The median size of all awards classified as net space research for purposes of this report was $68,000 in FY 1986 and $70,000 in FY 1995. This median reached a minimum during the intervening years. In comparison, the mean value of research grants awarded by the EPA was $116,000 per year,[10]

---

[10]Environmental Protection Agency, Office of Research and Development, presentation to the Board of Scientific Counselors meeting, October 1997.

TABLE 4.4 Selected Statistics on NASA Awards to Colleges and Universities, FY 1986-1995

Characteristics of NASA Procurement Awards	New and Continuing Awards				New Awards Only			
	FY 1986	FY 1989	FY 1992	FY 1995	FY 1986	FY 1989	FY 1992	FY 1995
All Awards (with training grants)								
Total number	2,814	3,739	4,799	5,069	755	1,161	1,270	1,419
Total value (million constant FY 1995 dollars)	362	549	718	826	70	101	102	170
Average value (thousand dollars)								
Mean	128	146	151	162	92	88	81	120
Median	67	60	54	55	48	48	38	48
Mode	24	22	24	22	24	22	24	22
Median duration (months)	25	24	25	35	11	11	12	13
Awards Classified as Net Space Research (without training grants)								
Total number	1,632	2,025	2,653	2,645	392	556	674	723
Total value (million constant FY 1995 dollars)	174	216	289	299	22	36	46	65
Average value (thousand dollars)								
Mean	107	107	109	113	76	78	74	90
Median	68	66	65	70	52	49	51	59
Mode	67	60	54	50	53	60	54	50
Median duration (months)	31	29	25	34	11	11	12	13
Deflator (FY 1995 = 100)	74.8	83.3	92.9	100.0	74.8	83.3	92.9	100.0

NOTE: Details may not add to totals because of rounding.

while the mean value of awards issued by the NSF in FY 1996 was $85,000 and the median was $52,000.[11]

• The most frequent size (i.e., modal value) for awards classified as net space research declined significantly over the decade, falling from about $67,000 in FY 1986 to $50,000 in FY 1995 (in constant FY 1995 dollars).

• The median size of OSS awards decreased by about $5,000 during the decade (see Figure 4.3).

• The median size of OES awards remained constant at about $80,000 during the decade.

• The median size of OLMSA awards increased significantly during the decade, from $69,000 in FY 1986 to $100,000 in FY 1995.

---

[11] National Science Foundation, Report on the NSF Merit Review System, FY 1996.

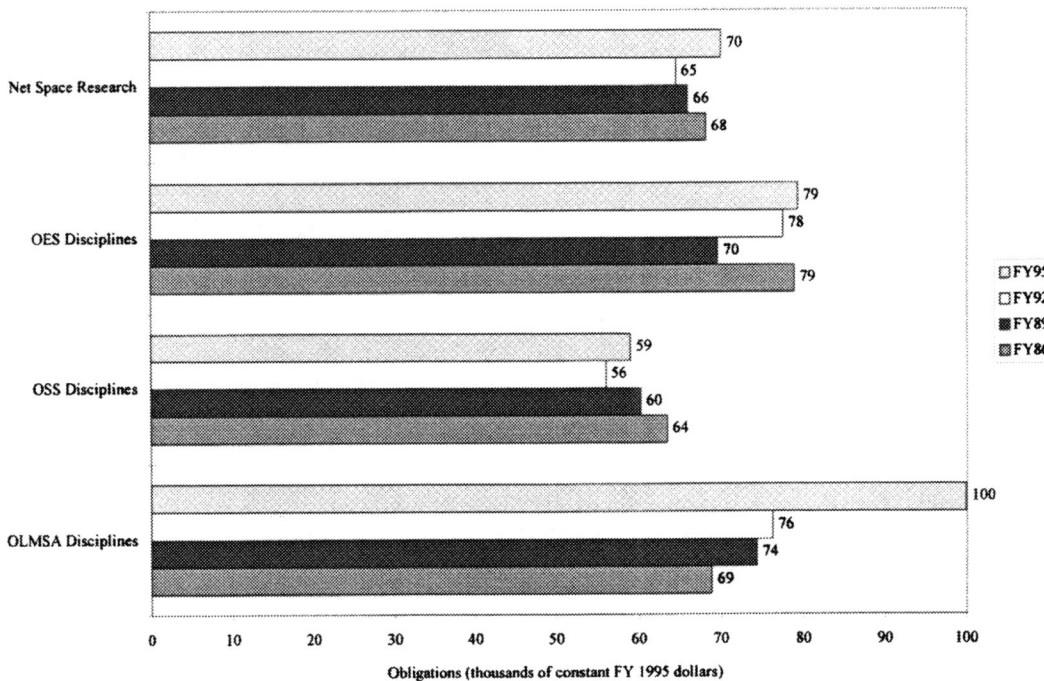

FIGURE 4.3 Median size of NASA awards for net space research and science program disciplines, FY 1986-1995.

4.5 CHARACTERISTICS OF GRANTS AT NASA FIELD CENTERS

There is a dearth of information about intramural research funding within NASA field centers or at JPL. NASA's accounting system is not structured to track the size of the in-house scientific staffs at NASA centers or the way in which these staffs are allocated between the functions of intramural research, science management, and other project or general management. Although there are procedures and policies for allowing NASA center scientists to participate in peer-reviewed NASA-funded science, there are no accompanying statistics indicating the extent (in terms of funding) to which these center scientists actually participate in the "research base."[12]

Data from the FY 1997 congressional budget justification books for NASA's three science program offices provide some information about budget allocations between the NASA centers and Headquarters. They do not indicate where the funds are actually being expended. Field centers do contract with universities and industry for some of their research.

Approximately 65 percent of OSS expenditures occur at the Goddard Space Flight Center and at JPL; approximately 85 percent of OES expenditures occur at Goddard Space Flight Center and at JPL,

---

[12]For many years, NASA has operated an integrated system of peer review for science investigator selection in response to formal notices of flight or other research opportunities from both universities and NASA centers. There are well-documented agency procedures for ensuring even-handed selection among competing proposals from investigators in the various research-performing institutions. Still, there has not been much systematic, and certainly not comprehensive, reporting of the outcomes of overall resource allocations (including budgets) between universities, NASA centers, and other external research-performing institutions.

and approximately 60 percent of OLMSA's expenditures are concentrated at Johnson Space Center, Kennedy Space Center, and Marshall Space Flight Center. The agency plans to implement a "full-cost" accounting system that should allow better estimates of intramural research costs and should aid in intramural versus extramural comparisons. Although JPL's status as an FFRDC precludes it from being treated identically with the NASA civil service centers, data concerning its in-house science programs could be collected and distributed like those of the field centers. In addition to JPL, several nonprofit institutions are involved in space research for which very few detailed data are reported. These include the Hubble Space Telescope Science Institute, the Smithsonian Astrophysics Observatory, and the Universities Space Research Association.

5

Science Community's Perceptions About the Research and Data Analysis Programs

The task group sought from the standing discipline committees of the Space Studies Board a sense of the research community's concerns about the R&DA programs.[1] Each committee was invited to comment on R&DA-related issues of innovation and technology insertion; smaller, faster, cheaper missions; "big science" versus science of opportunity; adequacy of facilities; and relative roles of NASA field centers and academia. Their responses are organized into four areas: (1) resource allocation; (2) technology, facilities, and infrastructure; (3) research grant management; and (4) maintenance of intellectual capital. The task group also notes whether or not its data validate committee perceptions.

5.1 RESOURCE ALLOCATION

Perception of Reduced Funding for Data Analysis

NASA has transferred much of the responsibility for data analysis from MO&DA accounts that had been part of flight projects to the general R&DA programs. This approach may allow flight projects to appear less costly because the expense no longer appears in the flight budget, but researchers note that these savings are illusory because the data must still be analyzed for the nation to benefit from the mission. Members of the scientific community argue that there is no evidence that R&DA accounts were increased to reflect increased responsibility for data analysis.

---

[1] These committees include the Committee on Astronomy and Astrophysics, Committee on Earth Studies, Committee on Microgravity Research, Committee on Planetary and Lunar Exploration, Committee on Solar and Space Physics–Committee on Solar-Terrestrial Research, and Committee on Space Biology and Medicine.

Task group finding. NASA budget and expenditure records did not allow the task group to determine whether equivalent funds for the new DA demands on R&DA programs were actually transferred with the tasks. The data do show (see Table 4.1, Chapter 4) that the fraction of R&A funding relative to NASA's science-related budget has decreased by approximately 35 percent over the past 8 years at the same time that MO&DA funding as a fraction of the total science budget fell about 30 percent. It is also clear that, although R&A has become a smaller fraction of the science enterprise, it has assumed roles that were once part of flight projects (see Chapter 3). These observations are consistent with scientists' sense of "shrinking" research dollars.

Perception of Excessive Concentration of R&DA Activities in NASA Field Centers and at the Jet Propulsion Laboratory

The external science community is concerned that a disproportionate fraction of R&DA research resides in the field centers and that the trend is toward increasing this fraction. Many scientists argue that field centers have a history of pulling work that was once in universities into the centers. The NRC report *Managing the Space Sciences* notes that field centers should maintain highly qualified, practicing scientists, but only insofar as the research these scientists conduct relates to the programs and projects the scientists are responsible for supporting.[2] However, some members of the external science community argue that prototype instruments, once the domain of universities and industry, are now thought to be developed disproportionately in field centers or at the Jet Propulsion Laboratory as "facility" instruments or as investments to capture future flight opportunities.

Despite the current emphasis on missions led by principal investigators (PIs), members of the science community note that it is very difficult for universities to acquire and maintain capabilities to build flight instruments. Moreover, some members note that instrument development teams and instrument design skills therefore cannot be maintained at universities without sufficient infrastructure, opportunities, and resources to build them. The university role in flight hardware development is important for the creation of new and innovative technologies and sensors and for the training of young scientists and engineers. For example, the value of developing these technologies within universities is recognized by the National Science Foundation through its instrument development postdoctoral program.

Task group finding. About 40 percent of NASA basic research (see Table 4.2) is performed in NASA field centers and at the Jet Propulsion Laboratory, and some fraction of the 30 percent of NASA basic research funds that go to industry likely is used for contractor support of field center research.

Perception About Opportunities in Universities for Instrument and Technology Development to Support Smaller and Shorter-Duration Missions

The NRC report *Managing the Space Sciences* describes the role of NASA field centers in managing complex missions. Scientists from the community at large value field center assistance with, for example, mission design, ground systems for mission operations, and facilities for integration and test of

---

[2]National Research Council, Space Studies Board, *Managing the Space Sciences,* National Academy Press, Washington, D.C., 1995, p. 44.

PI-supplied spacecraft.[3] Moreover, they recognize NASA's increased use of the PI mode as the management approach for Explorer-class missions.[4] Having said this, the task group fully endorses the assertion from *Managing the Space Sciences* "that scientific research should, for the most part, be conducted outside the agency."[5]

Task group finding. Contrary to the example in this perception, the task group's data do not show a decline in NASA awards to universities for instrument and spacecraft development. If anything, the percentage of NASA basic research awards to universities for these activities rose slightly over the past decade. The fraction of research funding going to field centers appears to have fallen (from about 45 percent, including intramural and FFRDCs) since 1991, while the fraction for universities has increased (from 21 to 26 percent) over the same period (see Table 4.2).

5.2 TECHNOLOGY, FACILITIES, AND INFRASTRUCTURE

Scientific discovery and productivity can depend as much on advances in technology as on developments in theory and analysis. Concerns expressed by members of the scientific community concentrate on a waning attention to facilities and infrastructure, poor linkage between science needs and technology development, access to NASA facilities, and peer review of technology development.

Perception of Reduced Support for Research Facilities in Academia

The Space Studies Board's discipline committees perceive that investments in research facilities in academia are being neglected. Facilities cannot be built under the smaller, faster, cheaper flight programs as they could be under the major flight programs of the past. Laboratories, observatories, and supporting infrastructure are funded only sporadically by R&DA programs.

Task group finding. During the decade preceding 1995, there was a decrease in funding for what the task group calls "support facilities" (technical and engineering support such as sounding rocket engineering support, flight management systems, and maintenance and operations of some specialized tracking facilities). Yet there was also a significant increase in what the task group has labeled "science facilities" (operation of NASA research support facilities such as the National Scientific Balloon Facility, the Poker Flats Rocket Range, the NASA Infrared Telescope Facility, and the MacDonald Laser Ranging Station) and in cumulative awards to universities (see Table 4.3).

The task group notes that these science and support facilities programs are difficult to track due to accounting changes and subsequent reallocations of funds between science and support programs. The task group finds that, when these facilities are considered together, funding for all facilities (science and support) has been stable or slightly increasing. Perhaps most importantly, the task group also finds that programs within these categories tend to support national facilities rather than facilities located on university or college campuses.

---

[3] National Research Council, Space Studies Board, *Managing the Space Sciences*, National Academy Press, Washington, D.C., 1995.

[4] National Research Council, Space Studies Board, *Assessment of Recent Changes in the Explorer Program*, National Academy Press, Washington, D.C., 1996, pp. 1-2.

[5] National Research Council, Space Studies Board, *Managing the Space Sciences*, National Academy Press, Washington, D.C., 1995, p. 44.

5.3 RESEARCH GRANT MANAGEMENT

Perception of Declining Success Rate and Decreasing Size of Awards

Investigators report that their success rate for winning grants is declining and that the size of their grants has decreased. Their response has been to submit more proposals. The nonproductive burden on the research community of writing and reviewing these proposals is large.

Task group finding. OSS estimates that the current success rate in space sciences is about 30 percent for all proposals and about 10 percent for new proposals. Based on task group discussions with NASA, OES estimates an overall success rate of approximately 25 to 30 percent, whereas OLMSA estimates that success rates range from 12 to 23 percent overall.[6] In other federal agencies such as the EPA, success rates are even lower, around 10 to 15 percent for the Science to Achieve Results (STAR) program,[7] and across NSF, success rates for FY 1996 awards were reported at 29 percent, having declined steadily over the past 5 years from 34 percent.[8] Although total funding of university grants for space research has increased, the number of awards to universities has also increased substantially (see Table 4.4). Consequently, the grant seen by the typical investigator has not increased: The median award size for so-called net space research increased only slightly from $68,000 to $70,000 (in constant 1995 dollars) per year over the decade, but the mode value—the size of the typical award—for net space research decreased by 25 percent in constant dollars over 10 years from $67,000 to $50,000. When the data are sorted by NASA program office, the task group finds that the problem of small and decreasing grant size was most severe in OSS. OLMSA, on the other hand, showed increases in award size over the period (see Figure 4.3). The task group's data do not permit it to determine the extent to which the increasing number of awards may be due in part to multiple awards to single investigators rather than to increasing numbers of investigators being sponsored.

5.4 INTELLECTUAL CAPITAL

Perception of Limited Opportunities to Gain Instrument Development Experience

Scientists from the external community explain that success with smaller, faster, cheaper missions entails the rapid incorporation of ideas and discoveries from past flights into new hardware and missions. This can happen only when many of the participating scientists and engineers have had experience in developing, testing, and flying new instruments. Senior scientists note that the number of investigators with instrument-building experience eroded during the period of large, but infrequent, missions.

Task group finding. Table 4.3 in Chapter 4 shows a significant increase in instrument development activity in universities, both in terms of funding level and as a fraction of the total awards to universities. This could be a reflection of the recent introduction of instrument incubator programs in some disciplines and a few other well-funded PI-class missions (e.g., the Far Ultraviolet Spectroscopic Explorer [FUSE]).

---

[6]Success rates for individual NASA proposal competitions will vary from one to another. The estimates for overall success rates are imprecise and informal.

[7]The U.S. EPA Office of Research and Development, National Center for Environmental Research and Quality Assurance.

[8]National Science Foundation, Report on the NSF Merit Review System, FY 1996.

5.5 OTHER PERCEPTIONS

There were other concerns raised by the Space Studies Board's discipline committee members that were not addressed by the task group's budget data. For example, scientists are concerned about shorter-duration missions, which may require investigators to seek funding for mission extensions through the R&DA programs. Members of the community also argue that NASA research announcements (NRAs) focus heavily on particular missions and data, thereby reducing a program's flexibility to fund cross-disciplinary research. In addition, some scientists perceive a lack of science-driven technology programs and reduced support for science facilities, such as high-altitude aircraft and suborbital platforms. Finally, members of the external community are concerned about difficulties in attracting the best talent to graduate programs. They sense that the decreasing employment opportunities for Ph.D. scientists limit the ability to attract innovative thinking to NASA-funded sciences. Although some of these concerns have been discussed earlier in this report, they are not particularly illuminated through analysis of funding data.

6

Findings and Recommendations

6.1 PRINCIPLES FOR STRATEGIC PLANNING

Having explored the role of R&DA in NASA's strategic planning in Section 3.2, the task group identifies five principal elements that figure prominently in the strategic planning process: (1) a foundation built on basic scientific goals, objectives, and key questions; (2) clear linkage between the science questions, supporting research and technology, advanced technology development, flight missions, research and data analysis, and the strategic plans; (3) use of the peer review process to determine the merit of the scientific goals, objectives, and key questions; (4) use of independent advisory bodies to regularly review the progress and future directions of the strategic plans; and (5) room for flexibility, innovation, and evolution within the strategic plans.

Some of NASA's science offices, for example the Office of Space Science, have been thorough in their involvement of the scientific community in the strategic planning process and in their use of independent peer review to evaluate strategic plans and their underlying science questions.[1] The extent to which NASA has devoted explicit attention to strategic and operational linkages between basic scientific goals, R&DA activities, and space flight missions is much less clear.

Finding: The task group finds that R&DA is not always thoroughly and explicitly integrated into the NASA enterprise strategic plans and that not all decisions about the direction of R&DA are made with a view toward achieving the goals of the strategies. The task group examined the trend and balance of R&DA budgets and found alarming results (Chapter 4, Sections 4.1 and 4.3); it questions whether these results are what NASA intends.

---

[1] For example, the NASA Space Science Advisory Committee held a workshop in May 1997 in Breckenridge, Colorado, to integrate science and technology "roadmaps" into a strategic plan for the Office of Space Science; see also letter sent by Space Studies Board Chair Claude Canizares to NASA Associate Administrator for Space Science Wesley Huntress on NASA's Office of Space Science draft strategic plan, August 27, 1997.

Recommendation 1: The task group recommends that each science program office at NASA do the following:

• Regularly evaluate the impact of R&DA on progress toward the goals of the strategic plans.

• Link NASA research announcements (NRAs) to addressing key scientific questions that can be related to the goals of these strategic plans.

• Regularly evaluate the balance between the funding allocations for flight programs and the R&DA required to support those programs (e.g., assess whether the current program can support R&DA for the International Space Station).

• Regularly evaluate the balance among various subelements of the R&DA program (e.g., theoretical investigations; new instrument development; exploratory or supporting ground-based and suborbital research; interpretation of data from individual or multiple space missions; management of data; support of U.S. investigators who participate in international missions; and education, outreach, and public information).

• Use broadly based, independent scientific peer review panels to define suitable metrics and review the agency's internal evaluations of balance.[2]

• Examine ways to maximize familiarity with contemporary advances and directions in science and technology in the process of managing R&DA, for example, via the appropriate use of rotators.[3]

6.2 INNOVATION AND INFRASTRUCTURE

Innovations often require state-of-the-art facilities. The task group found evidence of few mechanisms to provide this essential research infrastructure. Section 3.4 notes that the difficulties experienced by universities, in particular, in acquiring and maintaining infrastructure are exacerbated in the smaller, faster, cheaper program environment where certified facilities and state-of-the-art laboratories might remain the expected norm, but there is neither the time nor the money to develop them that there was in the era of large, longer-duration flight projects.

Finding: Although there are sporadic funding opportunities for research infrastructure, there is no systematic assessment of the state of the research infrastructure, nor are there coherent programs to address weaknesses in the infrastructure base (Section 5.2).

Recommendation 2: The task group recommends that NASA take the following actions on research infrastructure:

• Conduct an initial assessment of the need and potential for acquiring and sustaining infrastructure in universities and field centers.

---

[2]National Research Council, Space Studies Board, "On NASA Field Center Science and Scientists," letter to NASA Chief Scientist France Cordova, March 29, 1995; National Research Council, Space Studies Board and the Committee on Space Biology and Medicine, "On Peer Review in NASA Life Sciences Programs," letter to Dr. Joan Vernikos, director of NASA's Life Sciences Division, July 26, 1995; National Research Council, Space Studies Board, "On the Establishment of Science Institutes," letter to NASA Chief Scientist France Cordova, August 11, 1995.

[3]Federal agencies have used rotators—scientists from outside the federal government—for 1 to 2 years to participate in management of research programs. NASA has used interagency personnel appointments—visiting scientists administered by the Universities Space Research Association and JPL—as rotators to circulate new ideas and new individuals, on temporary appointments, into the agency system.

- Determine options for minimizing duplication of expensive research facilities.
- Evaluate the level of support for infrastructure in the context of the overall direction and plans for R&DA activities.
- Maximize the use of infrastructure by supporting partnering between universities and field centers.
- Explore approaches for providing peer review and oversight of infrastructure investments, which should include regular evaluation of a facility's role and contribution as a national academic resource, its degree of scientific and technical excellence, and its contribution to NASA's long-term technology planning and development.
- Institute periodic assessment of the research infrastructure in university and NASA field centers to ensure that the infrastructure is appropriate for current programs.

6.3 MANAGEMENT OF THE RESEARCH AND DATA ANALYSIS PROGRAMS

University investigators report increasing competition for fewer grants, a lower success rate for each competition, lower dollar awards for successful grants, and hence the need to secure a larger number of grants to maintain a viable research program. The effect has been to increase the number of proposals that each investigator prepares and the number of grants that each investigator manages. Inefficiencies associated with competing for, evaluating, and administering a larger number of smaller grants appear not only as increases in the time devoted to writing and reviewing proposals and reports but also as an unproductive fragmentation of the efforts of investigators. Most scientists view this fragmentation as a churning that may produce grants and papers but that compromises the focus essential for discovery or innovation.

The current system encourages investigators to assume more work than they can reasonably complete well. Because of the relatively low success rate of proposals, the need for several concurrent grants to maintain a viable research program, and the dire consequences for academic scientists of finding themselves without funds to support their students, most investigators propose more work than they could possibly undertake if all proposals were successful. Many scientists conservatively underestimate their eventual success, with the result that their graduate students' research may be inadequately supervised or similar work may be performed under more than one grant. Although neither outcome is unethical, neither is likely to produce seminal discoveries.

Finding: The median size of NASA research grants to universities decreased in constant FY 1995 dollars from $64,000 per year in FY 1986 to $59,000 in FY 1995 for the Office of Space Science disciplines, remained relatively flat at $79,000 for Earth science disciplines, and grew from $69,000 to $100,000 for life and microgravity science disciplines during the period from 1986 to 1995 (Section 4.4, Figure 4.3). (These award sizes compare to a median of $85,000 at the National Science Foundation and a mean of between $110,000 and $120,000 at the Environmental Protection Agency.) It is well known that a single researcher cannot support a salary and a graduate student at grant levels of $50,000 and that such researchers must seek additional grants to maintain a viable research program.

Recommendation 3: NASA should routinely examine the size and number of grants awarded to individual investigators to ensure that grant sizes are adequate to achieve the proposed research and that their number is consistent with the time commitments of each investigator. The differences in award sizes for the Offices of Space Science, Earth Science, and Life and Microgravity Science and Applications should be reconciled with program objectives, especially those for space sciences, which often are funded at levels of less than $50,000 to $60,000. Where warranted, actions should be taken to address the deficiencies.

6.4 PARTICIPATION IN THE RESEARCH AND DATA ANALYSIS PROGRAMS

The competition between academic research and research at NASA field centers is both intense and mutually supportive. In defining their appropriate roles, the task group fully endorses the following excerpt from the report of the Space Studies Board's Committee on the Future of Space Science:

> NASA requires in-house scientists for its space research, exploration, and technology programs. These scientists coordinate science and operations on larger missions, guide development and utilization of unique research facilities, assist outside scientists and technologists to effectively use NASA facilities or flight opportunities, and enable NASA to act as a "smart buyer." The number of in-house scientists should be determined by the extent of these support functions and not by a desire to exploit perceived flight opportunities. Space science leadership and the generation and testing of new ideas should be the domain of the broader scientific community, of which the NASA scientists are only a part. As noted earlier, the committee believes that scientific research should, for the most part, be conducted outside the agency.[4]

Within the context of R&DA programs, the task group wants to emphasize the value of PI-led instrument development projects within academia. These projects are often the incubators of the next generation of flight projects and play an essential role in preparing the next generation of investigators for the responsibility of leading costly spaceflight projects. The issues of relative funding allocations and responsibilities are addressed in Section 4.2 and in Sections 5.1 and 5.4, respectively.

Finding: The task group recognizes that university-based instrument development projects led by principal investigators (PIs) can provide important training and versatility for graduate students in NASA-funded sciences. Often, innovative instrument prototypes can be developed at a fraction of the cost of facility instruments, and the analysis of instrument data and the preparation of high-quality scientific results are closely coupled with understanding of and experience in the design of scientific instrumentation. However, although the university arena frequently offers these opportunities, the task group also recognizes that some research facilities do not offer training advantages, that the economies of scale for some facility development projects are high, and that support of nonuniversity, multiuser facilities is sometimes necessary.

Recommendation 4: NASA should preserve a mix of PI-university awards and nonuniversity funding for the development of technologies, instruments, and facilities. NASA should make these decisions within the agency's overall plan for R&DA activities (Recommendation 1), with sensitivity to the advantages of the academic environment but guided by peer review of scientific and technical merit.

---

[4]National Research Council, Space Studies Board, *Managing the Space Sciences,* National Academy Press, Washington, D.C., 1995, pp. 43-44.

6.5 CREATION OF INTELLECTUAL CAPITAL

Given the limited opportunities for growth in the space-related sciences, colleges and universities must develop the next generation of science and engineering leadership without training a large cohort of investigators who expect to find research positions that will not materialize. The many solutions being discussed in the literature range from "birth control" of Ph.D.s to laissez faire (i.e., students carrying total responsibility for assessing their chances of meaningful employment). The task group is most comfortable with the middle position of insisting on a sufficient breadth in graduate education so that students are prepared to follow jobs beyond the narrow disciplines of their thesis research.

Supporting first-year graduate students on research assistantships presents faculty advisers and their students with significant incentives to focus narrowly on a research problem from the very beginning of graduate education. The task group thinks that all of NASA's current graduate student support that falls under training grants is structured to encourage a relatively narrow focus on research. An alternative mode, the NIH and NSF training grant, emphasizes the quality and breadth of an academic plan and breadth of laboratory experiences.[5]

Finding: NASA's principal graduate student fellowship programs are all tied to student research interests or concentrations.

Recommendation 5: NASA should explore using training grants like those of the National Institutes of Health and the National Science Foundation for first-year graduate students as a possible alternative to supporting these students as research assistants or NASA fellows. These training grants should be designed to ensure breadth in graduate education and thereby may expand students' opportunities for employment within or beyond NASA-funded sciences.

6.6 ACCOUNTING AS A MANAGEMENT TOOL IN THE RESEARCH AND DATA ANALYSIS PROGRAMS

As noted in Chapter 4, the task group's review of NASA's R&DA programs was challenged by the limited systematic budgetary and expenditure data available about these programs. NASA is taking important first steps in this direction with full-cost accounting at its field centers, but much more than this is needed (see Section 4.5). There is a need for documentation and mappings between old account items and new items when change is necessary, for long-term tracking of classes of program support (e.g., instrument development, infrastructure, data analysis), for long-term tracking of allocations within each class for types of participants (e.g., field centers, universities, industry), and for openly reporting these budget and expenditure indicators each year.

---

[5]The Committee on Science, Engineering, and Public Policy recommended in its report *Reshaping the Graduate Education of Scientists and Engineers* (National Academy Press, Washington, D.C., 1995): "To produce scientists and engineers who are versatile, graduate programs should provide options that allow students to gain a wider variety of academic and other career skills," and "To foster versatility, government and other agents of financial assistance for graduate students should adjust their support mechanisms to include new 'education/training grants' that resemble the training grants now available in some federal agencies" (pp. 78-79). On undergraduate education, see *Shaping the Future: New Expectations for Undergraduate Education in Science, Mathematics, Engineering, and Technology,* a report on its Review of Undergraduate Education by the Advisory Committee to the National Science Foundation Directorate for Education and Human Resources, M.D. George, Chair of the Review Subcommittee, NSF 96-139, National Science Foundation, Washington, D.C., May 1996.

Care should be taken to implement this recommendation in its intended spirit—that is, to provide a management tool. The effort could be easily subverted to mask actual practices if demonstrating a desired outcome became the de facto norm. For example, science support by university or industry contractors performed on or near a field center might be classified as research in universities or in industry rather than by field centers, or a pass-through contract from a university to industry might be classified as university instrument building rather than industry instrument building.

Finding: NASA does not use the extended records of its budgets and expenditures as management tools to monitor the health of its R&A and DA programs. Moreover, the fragmented budget structure for R&DA makes it difficult for the scientific community to understand the content of the program and for NASA to explain the content to federal budget decision makers.

Recommendation 6: NASA's science offices should establish a uniform procedure for tracking budgets and expenditures by the class of activities and the types of organizations (including intramural and extramural laboratories, industry, and nonprofit entities) that are actually performing the work. These data should be gathered and reported annually and used to inform regular evaluations of R&DA activities (Recommendations 1 and 2). One approach would be to itemize the following elements in the budget: theoretical investigations; new instrument development; exploratory or supporting ground-based and suborbital research; interpretation of data from individual or multiple space missions; management of data; support of U.S. investigators who participate in international missions; and education, outreach, and public information. In addition, these data should be made publicly available and reported annually to the Office of Management and Budget and to Congress.

Appendixes

A

Sources of Data and Method of Development

This appendix summarizes the various sources of data used by the Task Group on Research and Analysis Programs to build a database on detailed National Aeronautics and Space Administration (NASA) procurement awards made during the 1980s and 1990s, and it describes the coding structure and analytical categories and techniques used to develop the data for the study reported on in Chapters 1 through 6 of this volume. A primary objective of this activity was to estimate the net space research[1] component of awards made by NASA to universities. Any necessary caveats that should be observed in using the data in the context of this particular study are also noted.

NASA BUDGET HISTORY—THE BROAD CONTEXT

NASA presents a very extensive budget submission to Congress each year in support of the president's overall budget request. These budget justifications include a great deal of financial information, as well as supporting narrative about goals, objectives, schedules, and accomplishments of the various program elements that constitute NASA's program budget. Obtaining a consistent picture of the budget over a long span of time can be quite difficult because of changes in NASA's program structure and even more importantly because of changes in the NASA organizational elements responsible for the advocacy and management of these programs.

The task group's best efforts at developing a consistent long-term history of the NASA budget are presented in Tables A.1 and A.2. Table A.1 summarizes the ground-based elements of the NASA budget—the principal focus of this study. Table A.2 summarizes the budget history for major NASA

---

[1] "Net space research" is a term used by the task group to indicate the research funded from an R&DA source as opposed to technology development, instrument development, and academic training that may be funded by other accounts.

TABLE A.1 FY 1981-1998 NASA Budgets: Ground-based Programs

Major Science-related Programs and Activities	Fiscal Year Obligations (in millions of current dollars)					
	1981	1982	1983	1984	1985	1986
Total Research and Analysis	231.4	208.4	236.8	246.9	260.5	268.7
P&A	37.7	22.9	28.5	35.9	39.9	49.0
Planetary	50.7	46.7	50.3	59.5	61.5	59.5
OSS subtotal	88.4	69.6	78.8	95.4	101.4	108.5
Life sciences	29.5	25.5	31.7	35.0	35.2	34.0
Microgravity science	9.5	12.0	13.1	11.0	11.7	12.1
OLMSA subtotal	39.0	37.5	44.8	46.0	46.9	46.1
Earth science	104.0	101.3	113.2	105.5	112.2	114.1
OES subtotal	104.0	101.3	113.2	105.5	112.2	114.1
Total Other Science Support						
OLMSA—aerospace medicine						
EOS science						
EOS mission science teams and guest investigators						
OES—Globe program						
Total Suborbital Program	39.9	43.8	48.1	52.5	58.7	59.9
P&A—suborbital	39.9	43.8	48.1	52.5	58.7	59.9
SOFIA						
Sounding rockets	25.0	24.4	27.0	27.8	25.7	23.1
Airborne research	4.5	17.5	17.6	18.9	22.0	25.0
Balloon program	1.4	1.9	3.5	5.8	6.8	6.1
Spartan program					4.3	5.7
OES—airborne science and applications						(25.0)
OES—UAVs						
Total MO&DA (adjusted)	100.7	87.9	99.9	111.5	165.2	213.7
P&A—excluding HST operations and servicing						
P&A—in budget books	38.9	45.3	61.4	68.1	109.1	111.7
HST operations and servicing included						
Planetary	61.8	42.6	38.5	43.4	56.1	67.0
OSS—adjusted for HST operations and servicing						
OSS—combined						
OSS—in budget book						
Earth science						35.0
Total EOS Data and Information System (EOSDIS)						
Total Supporting Infrastructure	4.5	4.3	7.5	8.9	16.2	17.6
OSSA Information Systems Office	4.5	4.3	7.5	8.9	16.2	17.6
P&A information systems						
CIESIN						
OES information systems						
High-performance computing and communications						
Socio-Economic Data Applications Center						
Landsat						
Data purchases						
Commercial remote sensing						
Advanced geostationary studies						

SOURCES OF DATA AND METHOD OF DEVELOPMENT 73

	1987	1988	1989	1990	1991	1992	1993	1994	1995	1996	1997	Approp. 1998
	301.9	307.9	357.0	398.9	409.0	395.6	410.4	427.1	429.5	426.4	405.4	378.7
	53.4	829	85.1	104.9	98.3	69.9	71.6	71.1	75.4	62.8		
	69.5	67.3	76.9	70.7	67.8	76.6	101.7	107.6	108.4	93.4		
	122.9	150.2	162.0	175.6	166.1	146.5	173.3	178.7	183.8	156.2	166.8	130.5
	41.8	38.4	38.2	44.4	56.3	62.9	52.9	55.1	50.7	55.2	58.0	53.7
	13.9	12.9	19.2	17.6	13.7	16.6	17.9	18.4	30.4	30.2	31.9	30.8
	55.7	51.3	57.4	62.0	70.0	79.5	70.8	73.5	81.1	85.4	89.9	84.5
	123.3	106.4	137.6	161.3	172.9	169.6	166.3	174.9	164.6	184.8	148.7	163.7
	123.3	106.4	137.6	161.3	172.9	169.6	166.3	174.9	164.6	184.8	148.7	163.7
									63.9	47.5	84.3	88.3
											3.8	7.5
									37.3	16.7	37.5	37.4
									26.6	30.8	41.8	45.9
											5.0	5.0
	79.1	66.5	68.4	72.1	75.2	80.4	85.5	94.7	93.2	115.3	79.2	105.9
	79.1	44.7	45.4	52.7	55.0	60.1	64.8	69.5	67.2	88.0	59.9	83.3
											21.3	45.8
	30.4	27.5	27.0	30.1	31.3	34.2	36.4	39.5	38.0	38.6	24.6	23.8
	35.6	7.3	9.8	10.7	11.5	12.0	13.0	13.6	13.2	33.4		
	7.9	9.9	8.6	11.9	12.2	13.9	15.4	16.4	16.0	16.0	14.0	13.7
	4.7											
	(35.6)	21.8	23.0	19.4	20.2	20.3	20.7	25.2	26.0	27.3	19.0	20.7
											0.3	1.9
	239.7	229.0	164.6	232.6	335.5	383.3	456.2	418.3	379.0	410.7	421.0	395.8
			53.9	76.6	125.9	167.5	198.8	190.0	190.7			
	131.0	140.5	142.4	215.7	311.9	375.2	415.5	405.2	427.4			
			88.5	139.1	186.0	207.7	216.7	215.2	236.7	190.7	213.7	180.4
	75.1	73.8	110.7	156.0	170.2	160.7	163.4	130.7	117.2			
					296.1	328.2	362.2	320.7	307.9	372.9	382.8	348.1
					482.1	535.9	578.9	535.9	544.6			
										563.6	596.5	528.5
	33.6	14.7	17.6	23.8	39.4	55.1	94.0	97.6	71.1	37.8	38.2	47.7
						36.0	77.7	130.7	188.2	220.6	247.2	234.6	209.9
	21.2	20.8	19.9	28.2	4.5	50.0	74.3	50.4	66.8	83.1	107.8	47.1
	21.2	20.8	19.9	28.2	4.5	4.5	4.5	4.5	4.5	4.5		
								25.0	26.5	26.1	25.9	
						25.0	18.0	5.0				
							11.2	11.2	9.7	9.6	8.5	4.3
									20.5	26.1	28.3	18.3
									6.0			
						7.5						
						13.0	15.6	3.2			50.0	
										17.0	19.0	21.5
											2.0	3.0

TABLE A.1 *Continued*

Major Science-related Programs and Activities	Fiscal Year Obligations (in millions of current dollars)					
	1981	1982	1983	1984	1985	1986
Total Science-related Technology Programs						
OSS—core technology program (not mission-specific)						
OLMSA—space product development						
Total Academic Programs						
Education						
Minority research and education						
Total	376.5	344.4	392.3	419.8	500.6	559.9
Recap in Constant 1995 Dollars						
Research and analysis	378.1	320.1	348.7	350.2	357.8	359.2
Other science support	0.0	0.0	0.0	0.0	0.0	0.0
Suborbital program	65.2	67.3	70.8	74.5	80.6	80.1
MO&DA	164.5	135.0	147.1	158.2	226.9	285.7
EOS Data and Information System (EOSDIS)						
Supporting infrastructure	7.4	6.6	11.0	12.6	22.3	23.5
Science-related technology programs						
Academic programs						
Total Ground-based Programs in 1995 Dollars	615.2	529.0	577.8	595.5	687.6	748.5
GDP Implicit Price Deflator (1995 = 100)	61.2	65.1	67.9	70.5	72.8	74.8

NOTE: CIESIN = Consortium for International Earth Science Information Network; EOS = Earth Observing System; GLOBE = Global Learning and Observations to Benefit the Environment; HST = Hubble Space Telescope; MO&DA = mission operations and data analysis; OES = Office of Earth Sciences; OLMSA = Office of Life and Microgravity Science and Applications; OSS = Office of Space Science; OSSA = Office of Space Science and Applications; P&A = physics and astronomy; SOFIA = Stratospheric Observatory for Infrared Astronomy.

SOURCES OF DATA AND METHOD OF DEVELOPMENT

	1987	1988	1989	1990	1991	1992	1993	1994	1995	1996	1997	Approp. 1998
										0.0	200.5	213.7
											187.5	200.8
											13.0	12.9
		(21.6)	(24.0)	37.5	55.1	66.8	92.9	85.5	106.2	109.9	120.4	120.0
					37.9	44.8	70.2	54.3	57.9	61.5	65.6	68.6
					17.2	22.0	22.7	31.2	48.3	48.8	54.8	51.4
	641.9	624.2	609.9	769.3	915.3	1053.8	1250.0	1264.2	1359.2	1440.1	1653.2	1559.4
	391.1	385.4	428.6	459.0	452.9	425.8	430.6	438.1	429.5	416.8	387.6	355.6
	0.0	0.0	0.0	0.0	0.0	0.0	0.0	0.0	63.9	46.4	80.6	82.9
	102.5	83.2	82.1	83.0	83.3	86.5	89.7	97.1	93.2	112.7	75.7	99.4
	310.5	286.6	197.6	267.7	371.5	412.6	478.7	429.0	379.0	401.5	402.5	371.6
					39.9	83.6	137.1	193.0	220.6	241.6	224.3	197.1
	27.5	26.0	23.9	32.5	5.0	53.8	78.0	51.7	66.8	81.2	103.1	44.2
											191.7	200.7
				43.2	61.0	71.9	97.5	87.7	106.2	107.4	115.1	112.7
	831.5	781.2	732.2	885.3	1013.6	1134.3	1311.6	1296.6	1359.2	1407.7	1580.5	1464.2
	77.2	79.9	83.3	86.9	90.3	92.9	95.3	97.5	100.0	102.3	104.6	106.5

TABLE A.2 FY 1981-1998 NASA Budgets: Major Flight Projects

Fiscal Year Obligations (in millions of current dollars)

Major Science-related Flight Projects	1981	1982	1983	1984	1985	1986	1987
Total Physics and Astronomy	188.8	162.8	251.3	330.2	364.1	259.3	202.2
SIRTF development (and ATD)							
ISPM development	28.0						
HST development	119.3	121.5	182.5	195.6	195.0	125.8	96.0
HST operations and servicing (adjustment)							
GRO development	8.2	8.0	34.5	85.9	117.2	85.3	50.5
AXAF development							
Global geospace science development							
TIMED development (and ATD)							
Payload and instrument development							
Relativity mission development (GP-B)						(7.5)	(9.0)
Explorer development	33.3	33.3	34.3	48.7	51.9	48.2	55.7
Total Planetary Exploration	63.1	120.7	97.6	114.5	173.3	227.1	214.6
Galileo development	63.1	115.7	91.6	79.5	58.8	64.2	71.2
Magellan development				29.0	92.5	120.3	97.3
Ulysses development (ISPM)		5.0	6.0	6.0	9.0	8.8	10.3
Mars Observer development					13.0	33.8	35.8
Mars Balloon Relay (Mars '94)							
Cassini development							
Discovery development							
Mars Surveyor program							
New Millennium ATD							
Origins ATD							
Exploration technology development							
Total Life and Microgravity Sciences and Applications	49.3	65.8	113.9	118.5	147.8	140.4	151.7
Lifesciences	12.7	14.0	24.0	23.0	27.1	32.1	30.0
Microgravity	9.2	4.2	8.9	14.6	15.3	18.9	33.4
Shuttle and Spacelab payloads	27.4	47.6	81.0	80.9	105.4	89.4	72.8
Space Station payloads and planning							15.5
Station research facilities (move to space station budget)							
Mission to Planet Earth (and precursors)	90.2	92.0	76.1	44.4	75.5	133.3	175.3
Landsat	88.5	81.9	58.4	16.8			
UARS		6.0	14.0	20.0	55.7	114.0	113.8
Topex							18.9
EOS							
Earth probes (including Scatterometer)					12.0	14.0	32.9
Space station attached payloads							
Payload and instrument development	1.7	4.1	3.7	7.6	7.8	5.3	9.7

SOURCES OF DATA AND METHOD OF DEVELOPMENT

1988	1989	1990	1991	1992	1993	1994	1995	1996	1997	Approp. 1998
233.0	406.8	452.1	505.6	661.1	674.6	707.3	737.4	667.9	476.9	573.1
								15.0	24.9	55.4
93.1	104.9	81.8								
	88.5	139.1	186.0	207.7	216.7	215.2	236.7	190.7	213.7	180.4
53.4	50.9	41.2	22.0							
	16.0	44.0	101.2	150.7	168.3	239.3	224.3	237.6	18.4	95.8
18.6	64.4	57.6	96.6	75.3	72.6	27.6	40.0			
								15.0	25.9	52.7
				118.3	74.2	59.5	66.0	25.9	16.9	18.0
(10.3)	(17.9)	(21.7)	(23.4)	(27.2)	27.0	42.4	50.0	51.5	59.6	57.3
67.9	82.1	88.4	99.8	109.1	115.8	123.3	120.4	132.2	117.5	113.5
186.6	229.0	164.2	235.8	296.9	208.5	413.0	454.6	405.6	241.4	266.7
51.9	73.4	17.1								
73.0	43.1									
7.8	10.3	14.3	2.8							
53.9	102.2	98.9	88.5	85.0						
		4.4	1.5	1.2	3.5	4.4				
		29.5	143.0	210.7	205.0	266.6	255.0	191.5	74.6	
						127.4	129.7	102.2	76.8	76.5
						14.6	59.4	111.9	90.0	145.2
							10.5			
										25.0
										20.0
150.3	173.0	226.1	261.5	276.9	336.7	434.0	379.7	312.8	137.0	109.3
33.8	40.9	61.7	81.1	94.7	81.1	131.7	89.8	54.4	39.4	34.8
49.8	56.4	84.3	88.6	104.2	156.0	156.6	97.1	76.3	73.4	69.6
47.8	67.7	75.1	88.8	78.0	94.1	108.7	102.3	53.6	24.2	4.9
18.9	8.0	5.0	3.0	(7.7)	5.5	37.0	90.5			
								128.5		
214.0	225.2	229.7	243.2	357.1	423.5	515.2	675.2	634.3	644.0	753.2
				(78.0)	25.0	(74.1)				
89.2	85.2	55.2	62.0							
74.5	83.0	84.8	80.4	65.0						
				176.4	263.7	392.9	574.1	554.2	582.2	704.6
22.6	10.6	13.6	51.7	77.8	99.4	96.4	81.6	80.1	61.8	48.6
27.7	46.4	76.1	49.1	37.9	35.4	25.9	19.5			

TABLE A.2 *Continued*

	Fiscal Year Obligations (in millions of current dollars)						
Major Science-related Flight Projects	1981	1982	1983	1984	1985	1986	1987
Total Space Station Research (budget restructured in FY 1999[a])							
Research projects							
Utilization support							
Mir support (including Mir research)							
Total Science-related Technology Programs	0.0	0.0	20.0	25.0	45.0	81.9	84.6
ACTS development			20.0	25.0	45.0	81.9	84.6
OSS Mission studies and technology development							
OSS Focused technology programs (mission-specific)							
OSS New Millennium program							
Total (including Space Station research facilities and focused technology programs)	391.4	441.3	558.9	632.6	805.7	842.0	828.4
Recap in Constant 1995 Dollars							
Physics and astronomy	308.6	250.1	370.2	468.5	500.2	346.5	262.1
Planetary exploration	103.1	185.4	143.8	162.5	238.1	303.5	278.1
Life and microgravity sciences and applications	80.6	101.1	167.8	168.1	203.0	187.6	196.6
Mission to Planet Earth (and precursors)	147.4	141.3	112.1	63.0	103.7	178.1	227.2
Science-related technology programs	0.0	0.0	29.5	35.5	61.8	109.4	109.6
Space station research facilities							
Total Major Flight Projects in 1995 dollars	639.7	678.0	823.4	897.6	1106.8	1125.1	1073.6
GDP Pride Deflator	61.2	65.1	67.9	70.5	72.8	74.8	77.2

NOTE: ACTS = Advanced Communications Technology Satellite; ATD = Advanced Technology Development; AXAF = Advanced X-ray Astrophysics Facility; EOS = Earth Observing System; GRO = Gamma Ray Observatory; HST = Hubble Space Telescope; ISPM = International Solar Polar Mission; TIMED = Thermospheric Ionosphere Mesosphere Energetics and Dynamics; UARS = Upper Atmosphere Research Satellite

[a]Data shown prior to FY 1997 are and were distributed in other budget elements.

SOURCES OF DATA AND METHOD OF DEVELOPMENT 79

	1988	1989	1990	1991	1992	1993	1994	1995	1996	1997	Approp. 1998
							(187.8)	(254.6)	(277.4)	82.2	95.3
							(43.1)	(112.8)	(131.3)	82.2	95.3
							(21.0)	(36.3)	(64.4)	(54.6)	(89)
							(123.7)	(105.5)	(81.7)	(59.3)	(37)
	75.6	74.8	60.0	34.0	18.7	4.0	3.0	2.3	26.7	72.3	210.4
	75.6	74.8	60.0	34.0	18.7	4.0	3.0	2.3			
									26.7		
										26.7	170.7
										45.6	39.7
	859.5	1,108.8	1,132.1	1,280.1	1,610.7	1,647.3	2,072.5	2,249.2	2,047.3	1,653.8	2,008.0
	291.5	488.4	520.2	559.6	712.0	708.1	725.5	737.2	652.9	456.0	538.1
	233.4	275.0	188.9	261.0	319.8	218.9	423.6	454.5	396.5	230.8	250.4
	188.0	207.7	260.2	289.5	298.2	353.4	445.2	379.6	305.8	131.0	102.6
	267.7	270.4	264.3	269.2	384.6	444.6	528.4	675.0	620.0	615.8	707.2
	94.6	89.8	69.0	37.6	20.1	4.2	3.1	2.3	26.1	69.1	197.6
									125.6	78.6	89.5
	1075.1	1331.3	1302.6	1416.9	1734.7	1729.2	2125.8	2248.6	2126.9	1581.3	1885.4
	79.9	83.3	86.9	90.3	92.9	95.3	97.5	100.0	102.3	104.6	106.5

flight projects. Because of the large year-to-year changes in budget resources required for the development of major spaceflight projects, it is essential to separate these projects to make any sense of the dollar trends in funding for the various NASA program activities. In developing Tables A.1 and A.2, the task group tracked through successive budget documents in order to tabulate the latest available "actual" budget for each program element. For purposes of comparison, any given budget usually shows budget data for 3 years: (1) the prior fiscal-year amount (the actual amount as recorded in the agency's financial accounts), (2) the current-year amount (an estimate of the ongoing year's activity at the time the budget request is submitted), and (3) the budget-request amount (which is frequently modified by Congress in the appropriations process). Tables A.1 and A.2 track the prior-year (or actual) amounts given in the various budget reports summarized in this study.

The budget history tables (and many of the other analytical summaries in this report) include constant-dollar series that were adjusted by the task group based on Office of Management and Budget (OMB) gross domestic product (GDP) implicit price deflators (see Table A.3), which can be found in the president's annual budget documents and are also available on the Government Printing Office Web site (<http://www.gpo.gov/su_docs/budget99/hist_wk1.html>).

DATA FROM THE NATIONAL SCIENCE FOUNDATION'S FEDERAL FUNDS ANNUAL SERIES—BROAD SECTORAL OVERVIEW

Table A.4 provides a broad historical perspective on performers of NASA-funded R&D. This summary is based on the National Science Foundation's (NSF's) federal funds series of statistics gathered and published each year by NSF's Science Resources Studies Division. These data are collected by NSF from each of the major R&D-performing agencies of the federal government and are statistical extracts or estimates derived from the annual budget documents. These agency statistics, collected for several decades on a consistent basis, are an important source of information about federal R&D programs. They provide detailed estimates of agency programs, based on the character of the work being supported (including estimates of the amounts provided for basic and applied research and for development activities). For the purposes of this report, the NSF data are significant because they provide the only regularly published data from NASA that contain a breakdown of the overall NASA R&D program by performing sector (e.g., industry, academic institutions, nonprofit organizations, and in-house NASA centers). In fact, this series is the only source of which the task group is aware that presents estimates of the dollar value of NASA intramural research.

One caveat regarding NSF federal funds data is that they represent agency estimates—in this case, NASA estimates—of program amounts directed to particular performers of R&D. The data received by NSF from NASA are not tied to specific contracts and grants and do not provide either a breakdown by major NASA field installations or separate identification of Jet Propulsion Laboratory R&D work, which is included in the general category of federally funded research and development centers (FFRDCs).

DATA AND ANALYTICAL CATEGORIES FOR SUMMARIZING NASA AWARDS TO UNIVERSITIES

NASA does not currently publish very much top-level *summary* data about the nature of its procurement awards for externally performed research, such as the average size of NASA contracts, their average duration, or their distribution by major fields of science. However, for many years NASA has

TABLE A.3 Deflators for Task Group on Research and Analysis Study

Fiscal Year	Base Year FY 1995		Base Year FY 1989	
	Index	Value of $300,000	Index	Value of $300,000
1981	61.2	184	73.5	220
1982	65.1	195	78.2	234
1983	67.9	204	81.5	245
1984	70.5	212	84.6	254
1985	72.8	218	87.4	262
1986	74.8	224	89.8	269
1987	77.2	232	92.7	278
1988	79.9	240	95.9	288
1989	83.3	250	100.0	300
1990	86.9	261	104.3	313
1991	90.3	271	108.4	325
1992	92.9	279	111.5	335
1993	95.3	286	114.4	343
1994	97.5	293	117.0	351
1995	100.0	300	120.0	360
1996	102.3	307	122.8	368
1997	104.6	314	125.5	377
1998 (estimated)	106.5	320	127.9	384
1999 (estimated)	108.7	326	130.5	391

NOTE: Based on gross domestic product deflator series, Council of Economic Advisors. FY 1996 to FY 1999 values downloaded from OMB Web site: <http://www.gpo.gov/su_docs/budget99/hist_wkl.html>. Deflators were updated using the base year 1992.

published annually in its Green Books[2] an exhaustive listing of all of its research and training awards to colleges and universities that provides very detailed information *at the level of specific contracts and grants*. In the most recent fiscal year for which data were available for this study, FY 1995, 8,141 active contracts and grants were awarded to academic institutions.[3]

NASA's Green Books contain a wealth of information at the level of individual contracts and grants, including a specific contract or grant award number, the name of the receiving institution, its location by state, a brief descriptive title of the effort covered, the period of performance, the amount of funding obligations in the current fiscal year and cumulatively over the life of the award, the name of the principal investigator(s), the names of the NASA contracting office and the NASA technical officer, and a standard government-wide designation (CASE code) of the appropriate field of science to which the award applies.

To have detailed statistics on NASA research contracts and grants for use in this report, the task group undertook a major data preparation job. Data from the Green Books were entered into a computerized database using a combination of optical character recognition (scanning) and manual data entry. The data represent 4 years taken at 3-year intervals, covering the decade from FY 1986 to FY 1995.

---

[2]The term "Green Books" refers to *NASA's University Program Active Projects* and *University Program Management Information System* prepared by NASA's Office of Human Resources and Education, Washington, D.C.

[3]The figure of 8,141 shown here differs from the figure of 5,069 shown in Table 4.4 owing to reporting procedures at NASA. When NASA publishes its awards data, the agency lists all active grants, including those that have not yet closed and carry no dollar obligations in the fiscal year.

TABLE A.4 Summary of NASA-funded Research and Development by Performer ($ thousands)

Performing Sector	FY 1981	FY 1982	FY 1983	FY 1984	FY 1985	FY 1986	FY 1987
Intramural	1,043,805	1,165,551	1,134,436	1,043,278	1,171,117	1,217,343	1,413,839
Basic	216,411	250,670	305,480	344,647	318,382	363,307	379,435
Applied	475,577	501,171	547,461	422,781	482,271	528,314	590,660
Development	351,817	413,710	281,495	275,850	370,464	325.722	443,744
Industrial Firms	2,096,328	1,432,593	901,847	1,120,182	1,311,876	1,276,774	1,479,327
Basic	160,722	118,961	114,551	198,387	181,273	280,517	228,008
Applied	318,502	272,114	251,327	399,380	381,421	430,838	482,244
Development	1,617,104	1,041,518	535,969	522,415	749,182	565,419	769,075
Universities and Colleges	171,308	185,630	189,357	203,846	237,260	254,027	293,644
Basic	124,418	125,876	140,081	148,442	176,886	182,928	220,060
Applied	32,734	29,976	29,600	28,445	36,535	41,906	42,964
Development	14,156	29,778	19,676	26,959	23,839	29,193	30,620
Nonprofit Organizations	98,830	104,511	97,289	89,419	82,167	101,127	102,529
Basic	14,770	18,191	22,571	23,875	18,570	27,102	27,478
Applied	24,346	29,590	33,954	30,313	25,472	34,080	34,040
Development	59,714	56,730	40,764	35,231	38,125	39,945	41,011
FFRDCs—Universities	79,008	182,519	305,120	350,255	512,366	542,407	475,303
Basic	11,775	18,409	30,626	35,183	51,500	54,585	153,149
Applied	23,647	36,876	61,099	70,269	102,756	108,864	98,287
Development	43,586	127,234	213,395	244,803	358,110	378,958	223,867
FFRDCs—Nonprofit	650	417	495	399	744	589	565
Basic	562	410	203	179	686	336	170
Applied	38	4	264	196	53	227	232
Development	50	3	28	24	5	26	163
TOTAL NASA *(thousand dollars)*	3,489,929	3,071,221	2,628,544	2,807,379	3,315,530	3,392,267	3,765,207
Basic	528,658	532,517	613,512	750,713	747,297	908,775	1,008,300
Applied	874,844	869,731	923,705	951,384	1,028,508	1,144,229	1,248,427
Development	2,086,427	1,668,973	1,091,327	1,105,282	1,539,725	1,339,263	1,508,480
Percentage Distribution							
Basic	15.1	17.3	23.3	26.7	22.5	26.8	26.8
Applied	25.1	28.3	35.1	33.9	31.0	33.7	33.2
Development	59.8	54.3	41.5	39.4	46.4	39.5	40.1
Total	100.0	100.0	100.0	100.0	100.0	100.0	100.0

NOTE: Updated from *Trends in the Structure of Federal Science Support*, Federal Coordinating Council for Science, Engineering, and Technology, Washington, D.C., December 1992.
[a]Estimate.
SOURCE: *Federal Funds for Research and Development*: Fiscal Years 1994, 1995, and 1996, Vol. 44, Detailed Statistical Tables, NSF 97-302, National Science Foundation, Washington, D.C., 1997.

SOURCES OF DATA AND METHOD OF DEVELOPMENT 83

FY 1988	FY 1989	FY 1990	FY 1991	FY 1992	FY 1993	FY 1994	FY 1995	FY 1996[a]	FY 1997[a]
1,335,244	1,733,436	1,968,411	2,112,018	2,248,412	2,333,246	2,271,257	2,253,736	2,403.417	2,271,038
343,494	453,571	462,326	443,568	493,224	531,300	490,977	494,548	502,309	471,238
487,057	621,095	585,278	689,997	624,747	718,691	669,163	516,974	568,085	558,897
504,693	658,770	920,807	978,453	1,130,441	1,083,255	1,111,117	1,242,214	1,333,023	1,240,903
1,961,867	2,425,921	3,284,775	3,666,972	3,765,022	4,112,193	4,304,780	4,686,599	5,258.376	5,062,671
292,129	420,412	517,124	530,970	476,767	505,806	706,865	636,359	666,222	614,453
531,576	584,723	562,848	673,284	577,976	737,054	914,794	1,249,402	1,418,350	1,484,756
1,138,162	1,420,786	2,204,803	2,462,718	2,710,279	2,869,333	2,683,121	2,800,838	3,173,804	2,963.462
337,930	433,665	470,746	533,728	586,299	613,742	520,988	708,064	708,064	708,064
254,936	298,712	307,909	357,838	402,370	420,977	421,481	480,168	480,168	480,168
54,749	88,416	113,906	116,576	109,412	117,998	13,389	107,421	107,421	107,421
28,245	46,537	48,931	59,314	74,517	74,767	86,118	120,475	120,475	120,475
112,927	149,406	168,154	212,892	265,181	280,276	255,670	269,069	286,949	274,716
29,022	39,264	42,123	50,042	60,192	62,909	60,605	59,733	60,554	56,317
31,789	40,483	36,638	48,706	51,638	61,111	57,912	60,810	69,439	71,152
52,116	69,659	89,393	114,144	153,351	156,256	137,153	148,526	156,956	147,247
559,567	630,070	619,257	736,342	791,023	685,258	771,176	1,043,913	903,310	803,894
183,777	197,589	298,324	317,164	333,519	308,077	259,528	301,133	294,301	257,229
106,962	121,724	119,776	132,384	117,709	107,856	88,333	120,780	95,997	89,618
268,828	310,757	201,157	286,794	339,795	269,325	423,315	622,000	513,012	457,047
790	3,083	2,402	1,940	2,205	2,198	7,022	3,771	3,771	3,771
384	1,856	1,530	1,131	1,269	1,176	1,988	1,292	1,255	1,267
208	171	127	112	49	78	1,022	559	611	629
198	1,056	745	697	887	944	4,012	1,920	1,905	1,875
4,308,325	5,375,581	6,513,745	7,263,892	7,658,142	8,026,913	8,130,893	8,965,152	9,563,887	9,124,154
1,103,742	1,411,404	1,629,336	1,700,713	1,767,341	1,830,245	1,941,444	1,973,233	2,004,809	1,880,672
1,212,341	1,456,612	1,418,573	1,661,059	1,481,531	1,742,788	1,744,613	2,055,946	2,259,903	2,312,473
1,992,242	2,507,565	3,465,836	3,902,120	4,409,270	4,453,880	4,444,836	4,935,973	5,299,175	4,931,009
25.6	26.3	25.0	23.4	23.1	22.8	23.9	22.0	21.0	20.6
28.1	27.1	21.8	22.9	19.3	21.7	21.5	22.9	23.6	25.3
46.2	46.6	53.2	53.7	57.6	55.5	54.7	55.1	55.4	54.0
100.0	100.0	100.0	100.0	100.0	100.0	100.0	100.0	100.0	100.0

Because of time constraints, the amount of manual effort required, and limitations in applying optical character recognition techniques to the hard-copy reports in its possession, the task group was not able to collect all of the Green Book data elements associated with each procurement award in the database it constructed. For example, technical descriptions were collected only for the very largest awards (i.e., those with obligations of $300,000 or more in any of the 4 years examined for the study).

Two of the task group's objectives in developing this database were to categorize the awards by science discipline and to estimate the research component of NASA awards flowing into the academic sector. To achieve these objectives, a framework was needed for allocating data in the newly constructed database. Table A.5 summarizes the analytical categories and coding structure used for this purpose. Under the column heading "Science Programs" in Table A.5 are four categories of space-science-related activities that account for much of NASA science funding at universities and colleges. The task group attempted to specifically identify all research contracts and grants, by title, for the largest of the awards (coded as RES in its database). The names of technical officers were also useful in this process. Specific contracts for hardware design and development were grouped based on technical

TABLE A.5 Analytical Categories for Summarizing NASA Awards to Universities

Categories of University Activity Supporting NASA Missions	Classification Codes	Comments
Science Programs		
Major research grants (>$300,000 in at least 1 year)	RES	Classified on the basis of technical descriptions
Smaller research grants (<$300,000 in all years)	RIS	Residual smaller grants not otherwise assigned to specific categories
Instrument design and development	IDD	Includes flight instruments and advanced technology development
Spacecraft design and development	SDD	Complete systems (e.g., GP-B, EUVE)
Technology Programs		
Technology development and application	TECH	Classified on the basis of technical descriptions
Technology transfer and commercialization	TTXR	Includes facilities established specifically for technology transfer
Educational Programs (including outreach)		
National Space Grant College awards	NSGC	Program established by Congress in 1988
Training grants	NGT	All grants or contracts with NGT prefix (except space grants)
Other educational and human resource development	EDU	Classified on the basis of technical descriptions
Infrastructure and Program Support		
Operation of NASA research support facilities	OPS	An example is the Poker Flats sounding range
Technical and engineering support	SUP	Classified on the basis of technical descriptions
Centers of excellence (institutional capabilities)	CENX	Variety of facilities sponsored by NASA centers and offices to serve specific programmatic purposes

descriptions and then were subdivided into two categories, instrument design and development (coded as IDD) and spacecraft development (coded as SDD). These hardware contracts tend to be large awards relative to awards for performance of ground-based research and analysis; they are of interest in their own right as an illustration of the scope and variety of science-related activities carried out for NASA by the universities. Smaller research awards, of which there are literally thousands, could not be identified separately because of time and data constraints. Instead, these were estimated as the residual category after all of the other nonresearch-related activities were removed; these awards were coded as RIS in the database.

The two categories in Table A.5 summarized under "Technology Programs" generally cover NASA programs in aeronautical research as well as activities focused specifically on transferring aerospace technologies to commercial and other non-NASA users. The larger awards (coded as TECH or TTXR) were classified specifically on the basis of technical descriptions. Smaller awards (not coded separately) in technology-related activities include all NASA contracts and grants that were classified as engineering, mathematics, or computer science on the basis of CASE codes assigned by NASA program and procurement staff as reported in NASA's Green Books.

Awards in two of the three categories under "Educational Programs" in Table A.5 were classified on the basis of technical descriptions. These include all National Space Grant College awards (coded NSGC) and all other education-related activities not specifically tagged as training grants in the NASA procurement system (coded EDU). NASA training grants can be identified easily in the Green Books by the prefix NGT that appears as the first three letters of the NASA contract or grant number (coded simply as NGT by the task group).

The final set of categories in Table A.5 summarizes all NASA technical support and infrastructure activities carried out in the academic sector. All of these awards were classified by the task group on the basis of technical descriptions. For example, among the variety of program support activities that universities provide for NASA in the science arena are operation of the Poker Flats sounding rocket range in Alaska and, until fairly recently, operation of the NASA High Altitude Balloon Facility in Texas (all such operational contracts are coded as OPS in the database). Universities sometimes perform technical or engineering support functions for NASA; examples include editorial support for one of the NASA Headquarters program offices or the processing of synthetic aperture radar data for another office (all such activities are coded as SUP in the task group's database). The "Centers of Excellence" category is a catchall grouping established by the task group for specific contracts to universities in which the technical description implies that the effort is being funded to provide an ongoing institutional capability (coded as CENX in the task group's database to reflect the rationale that these activities are supported by NASA in order to establish and maintain a particular "center of excellence" for a specific programmatic purpose). These various broad infrastructure and program support activities relate to NASA's science, technology development, and educational support missions.

ESTIMATE OF NET SPACE RESEARCH COMPONENT OF NASA RESEARCH AND ANALYSIS AWARDS

The task group used the coding structure described above to categorize the large number and variety of contracts and grants awarded by NASA to universities and colleges to implement the agency's broad array of programmatic responsibilities. This categorization offers a means for characterizing the uses of funds awarded in all NASA contracts and grants to the academic sector. This approach also makes it possible to focus on the specific category of research contracts and grants of most interest to the task group's study of R&A programs, namely, the "net" space research component of NASA-sponsored

TABLE A.6 Estimation of the Space Research Component of NASA Awards to Universities

Categories of University Programs Supporting NASA Missions	Classification of Space Science Code(s)	Larger Awards (>$300,000)			
		Number of Awards			
		FY 1986	FY 1989	FY 1992	FY 1995
Research Contracts and Grants					
OLMSA disciplines	RES/LS	4	6	8	12
	RES/MGS	3	2	8	5
Subtotal OLMSA disciplines		7	8	16	17
OSS disciplines	RES/AA	11	16	24	22
	RES/SSP	13	16	18	13
	RES/LPX	5	10	9	4
Subtotal OSS disciplines		29	42	51	39
OES disciplines	RES/ES	18	12	26	48
Subtotal OES disciplines		18	12	26	48
Subtotal above: net space research		54	62	93	104
Percentage of total NASA awards		1.9	1.7	1.9	2.1
Other Space Science Activities					
Instrument design and development	IDD/Various	25	39	42	38
Spacecraft design and development	SDD/Various	3	3	4	4
Operation of science facilities	OPS/Various	2	2	9	9
Operation of support facilities	SUP/Various	7	11	5	6
Centers of excellence	CENX/Various	2	1	3	6
Subtotal (includes net space research—above)		93	118	156	167
Other NASA Activities					
Training grants	NGT	9	8	14	23
National Space Grant College awards	NSGC	0	0	20	20
Other education programs	EDU	3	3	11	29
Centers of excellence	CENX/NSS[a]	2	17	26	37
Technology programs	TECH	8	37	36	31
Technology transfer programs	TTXR	7	18	17	19
Balance of NASA University Awards					
Total university awards		122	201	280	326

[a]NSS = not space science.

SOURCES OF DATA AND METHOD OF DEVELOPMENT

CASE Field	Smaller Awards (<$300,000) Number				Consolidated Awards Number			
	FY 1986	FY 1989	FY 1992	FY 1995	FY 1986	FY 1989	FY 1992	FY 1995
Agricultural science	13	8	11	4				
Biological science	119	113	141	137				
Environmental biology	15	23	26	27				
Life sciences (not elsewhere classified)	33	24	34	84				
Medical sciences	38	52	55	47				
	218	220	267	299	225	228	283	316
Astronomy	375	484	731	742				
Chemistry	76	81	72	65				
Physics	231	342	393	383				
Physical science (not elsewhere classified)	97	143	242	187				
	779	1,050	1,438	1,377	808	1,092	1,489	1,416
Atmospheric science	266	301	371	375				
Environmental science (not elsewhere classified)	61	88	152	188				
Geological science	219	250	252	228				
Oceanography	34	53	80	74				
	580	692	855	865	598	704	881	913
	1,577	1,962	2,560	2,541	1,631	2,024	2,653	2,645
	56.0	52.5	53.3	50.1	58.0	54.1	55.3	52.2
					25	39	42	38
					3	3	4	4
					2	2	9	9
					7	11	5	6
					2	1	3	6
					1,670	2,080	2,716	2,708
Various	217	449	729	933	226	457	743	956
			30	30	0	0	50	50
					3	3	11	29
					2	17	26	37
Sum engineer, mathematics, computer science	820	1,035	1,092	1,073	828	1,072	1,128	1,104
					7	18	17	19
Not distributed	78	92	138	196	78	92	138	196
	2,692	3,538	4,519	4,743	2,814	3,739	4,799	5,069

space science in the academic sector. Since there is no means of directly measuring the R&A and DA (data analysis) activities being carried out under NASA contracts and grants, the subset of such awards was estimated by excluding awards for all activities related to other NASA (nonscience) programs and missions. Although this approach does not yield a perfect measure, it represents the only workable means available to the task group to develop a reasonable basis for assessing net space research activities of particular interest to this study. The approach, which is reproducible and has the additional merit of providing a consistent basis for making a time-based assessment of net space research activities, was applied to the historical data for all 4 fiscal years for which detailed NASA award statistics were collected—FY 1986, FY 1989, FY 1992, and FY 1995.

Assignment of Awards to NASA Science Disciplines

The two-stage process used by the task group to assign larger and then smaller NASA contracts and grants to the general analytical categories described above was also used to develop historical statistics relevant to the major NASA science disciplines and the three NASA science program offices—the Office of Earth Science (OES), the Office of Life and Microgravity Sciences and Applications (OLMSA), and the Office of Space Science (OSS)—currently responsible for science management of the major disciplines in NASA Headquarters. These allocations are perhaps best explained by referring to one of the worksheets used to develop the award count statistics for the 4 fiscal years covered by the study. Table A.6 provides this summary.

As Table A.6 shows, the count of all NASA awards to colleges and universities was distributed by the task group into analytical categories under the headings "Research Contracts and Grants," "Other Space Science Activities," and "Other NASA Activities." The larger awards (those totaling >$300,000 in any of the 4 fiscal years) were assigned classification codes as described in the preceding paragraphs. Appended to these classification codes is an additional science discipline code. Each of the large research awards was categorized by major NASA science discipline and by program office as follows: all awards classified as life science (LS) or microgravity science (MGS) were assigned to the current OLMSA; all awards classified as astronomy and astrophysics (AA), space and solar physics (SSP), or lunar and planetary exploration (LPX) were assigned to the current OSS; and all large research awards categorized as Earth science (ES) were assigned to the OES. The resulting allocations are subtotaled in Table A.6 to provide a comprehensive total of all the "net" space research contracts and grants awarded by NASA to universities for fiscal years 1986, 1989, 1992, and 1995. In summary, this is a direct allocation of all the larger awards based on examination of the technical descriptions for each specific award.

For smaller awards, it was simply not possible to apply this very time-consuming award-by-award classification scheme. Allocation of the thousands of smaller awards was achieved by assigning each award to the corresponding NASA science program office on the basis of the CASE science fields assigned by NASA personnel responsible for the database used to produce the Green Books. The task group notes that these allocations are somewhat arbitrary, but if the CASE codes are reasonably accurate, this approach should provide a reasonable basis for assigning awards to each of NASA's three science program offices. Finally, as described in the previous section, all smaller awards with CASE field codes in engineering, mathematics, or computer science were assigned to NASA's technology development mission and excluded from the estimate of the net space research component of NASA academic funding.

The consolidation of estimates in Table A.6 sums both the larger and smaller awards listed. The interesting substantive result of this detailed estimating procedure is that more than half of all NASA contracts and grants to universities fall within the broad category of net space research, the principal focus of the task group's study. This estimating approach was applied consistently for each of the 4 fiscal years for which detailed data were collected. These data suggest that over the decade from FY 1986 to 1995, the total number of NASA awards to universities and colleges for the performance of space research increased by about 1,000—from 1,631 awards in FY 1986 to 2,645 awards in FY 1995. Awards for purposes other than research increased more rapidly during the decade, with the result that the proportion of awards for research declined from about 58 percent of the total in FY 1986 to 52 percent in FY 1995.

Caveats and Additional Observations

Descriptive Statistics on NASA Contract and Grant Awards

Many task group members were concerned that the use of simple average statistics (especially the use of mean values) would not give an accurate sense of the variability or the typical value of award sizes, especially since these distributions are known to include relatively small numbers of very large awards and relatively large numbers of small-dollar-value awards. To address similar concerns, statistics on new versus continuing awards were tabulated to allow for the fact that some awards continue for very long periods whereas other do not. During a period in which the total number of awards is increasing, the average duration of awards tends to decline because of the varying proportions of new versus continuing awards.

Coded Large NASA Awards

In the process of developing data for analysis, the task group created a lengthy listing of large NASA awards that were then classified on an individual basis for purposes of this study. This listing, sorted by the major classification codes used to generate many of the statistical series reported in the text of this report, provides a basis for assessing the validity of the coding scheme used by the task group both for correlating activity types with the various awards and for assigning them to the major NASA science disciplines.

B

Overview of NASA Structure and Budget

For policy makers, budget analysts, recipients of research and data analysis (R&DA) grants, students, and interested analysts who may not be familiar with the overall structure and budget of the National Aeronautics and Space Administration (NASA), this appendix provides the broad agency context for the detailed discussions of R&DA in the report. Figure B.1 depicts the current organizational structure of NASA; Figure B.2 shows the distribution of the overall NASA budget for FY 1997; and Figure B.3 shows, more specifically, the budget for NASA science-related programs and activities broken down by the categories presented in the report (see Chapter 4, Box 4.1).

Research and Data Analysis (R&DA) activities are managed by the Office of Space Science, for the Space Science Enterprise; the Office of Earth Science, for the Earth Science Enterprise; and the Office of Life and Microgravity Sciences and Applications and the Office of Space Flight, for the Human Exploration and Development of Space Enterprise. Graduate student fellowships are administered by the Office of Human Resources and Education and the Equal Opportunities Programs. Aeronautical centers are administered by the Office of Aeronautics and Space Technology Transportation.

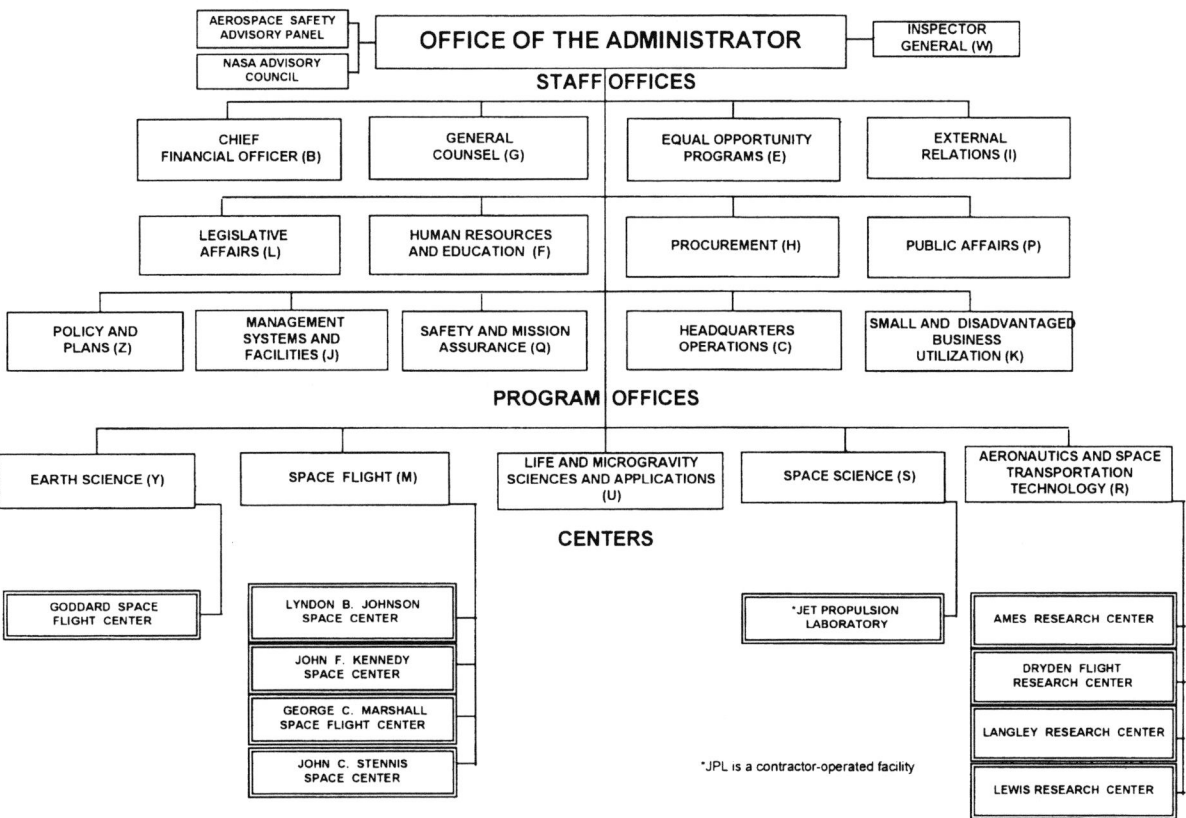

FIGURE B.1 Overview of NASA organizational structure, 1998.
SOURCE: National Aeronautics and Space Administration Headquarters.

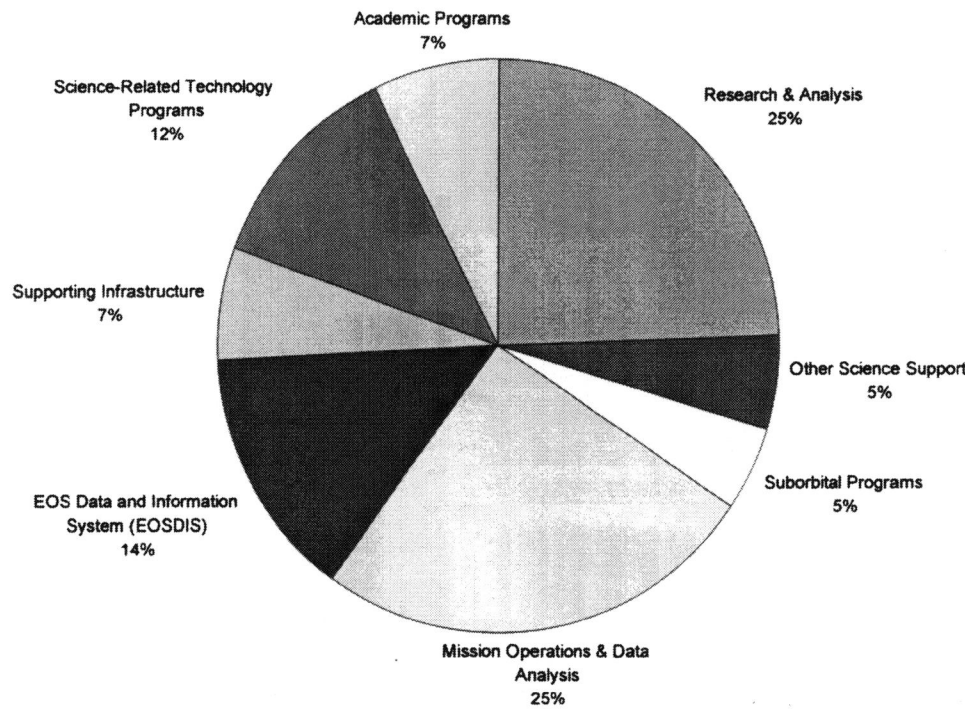

FIGURE B.2 Distribution of overall NASA budget, FY 1997.

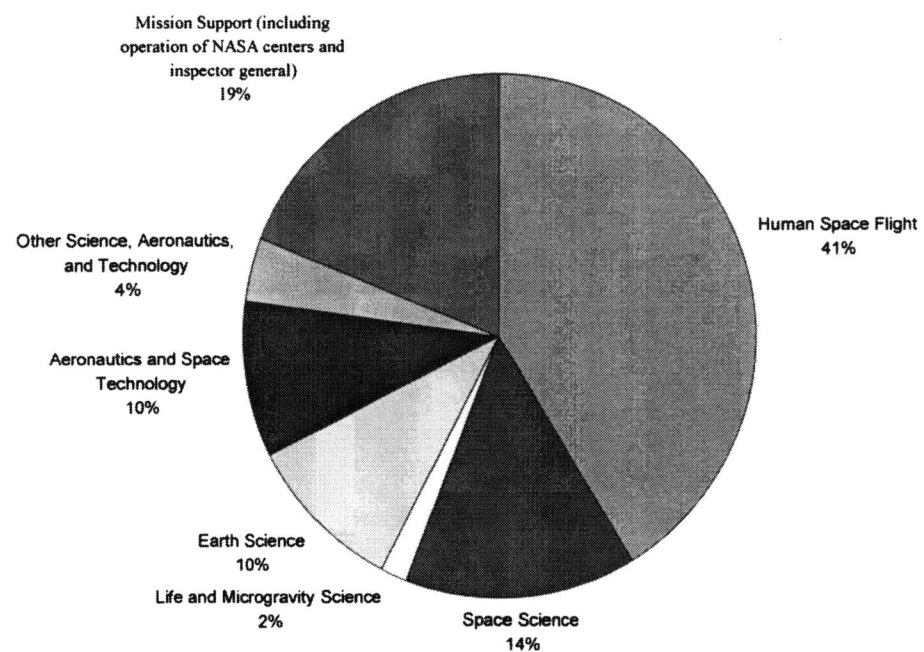

FIGURE B.3 NASA budget for science-related programs and activities, FY 1997 (constant FY 1995 dollars).

C

Acronyms and Abbreviations

AA	astronomy and astrophysics
ADEOS	Advanced Earth Observing Satellite
AgRISTARS	Agriculture and Resources Inventory Surveys Through Aerospace Remote Sensing
AHST	advanced human support technology
AO	announcement of opportunity
ATD	advanced technology development
AURA	Association of Universities for Research in Astronomy
AXAF	Advanced X-ray Astrophysics Facility
BR&C	biomedical research and countermeasures
CFC	chlorofluorocarbon
CT	computed tomography
DA	data analysis
ENSO	El Niño-Southern Oscillation
EOS	Earth Observing System
EOSDIS	Earth Observing System Data and Information System
EPA	Environmental Protection Agency
ES	Earth science
ESA	European Space Agency
ESSP	Earth System Science Pathfinder (program)
EUV	extreme ultraviolet
FFRDC	federally funded research and development center
FUSE	Far Ultraviolet Spectroscopic Explorer
FY	fiscal year
GB&E	gravitational biology and ecology

GEWEX	Global Energy and Water Cycle Experiment
GLOBE	Global Learning and Observations to Benefit the Environment (program)
GRO	Gamma Ray Observatory
GSRP	Graduate Student Research Program (NASA)
HEDS	human exploration and development of space
HRTEM	high-resolution transmission electron miscroscopy
HSCT	high-speed civil transport
HST	Hubble Space Telescope
IRTF	Infrared Telescope Facility
ISEE	International Sun-Earth Explorer
ISS	International Space Station
JPL	Jet Propulsion Laboratory
LACIE	Large Area Crop Inventory Experiment
Landsat	land remote-sensing satellite
LPX	lunar and planetary exploration
LS	life science
MGS	microgravity science
MIDEX	Medium Explorer
MO&DA	Mission operations and data analysis
MR	magnetic resonance
MURED	Minority University Research and Education Division (NASA)
NASA	National Aeronautics and Space Administration
NCEP	National Centers for Environmental Protection
NEAR	Near-Earth Asteroid Rendezvous
NICMOS	Near Infrared Camera and Multi-object Spectrometer
NIH	National Institutes of Health
NIST	National Institute of Standards and Technology
NOAA	National Oceanic and Atmospheric Administration
NRA	NASA research announcement
NRC	National Research Council
NSCAT	NASA Scatterometer
NSF	National Science Foundation
NSIPP	NASA Seasonal to Interannual Prediction Project
OES	Office of Earth Science (NASA)
OLMSA	Office of Life and Microgravity Science and Applications (NASA)
OMB	Office of Management and Budget
OMTPE	Office of Mission to Planet Earth (NASA)
OSAT	Office of Space Access and Technology (NASA)
OSF	Office of Space Flight (NASA)
OSS	Office of Space Science (NASA)
P&A	physics and astronomy
PAH	polycyclic aromatic hydrocarbon
PI	principal investigator
PIDDP	Planetary Instrument Definition and Development Program
R&A	research and analysis
R&DA	research and data analysis

SAR	synthetic aperture radar
SEASAT	Sea Satellite
SFC	smaller, faster, cheaper
SGER	Small Grants for Exploratory Research (NSF program)
S-I	seasonal-to-interannual
SMEX	Small Explorer
SMMR	Scanning Multichannel Microwave Radiometer
SOFIA	Stratospheric Observatory for Infrared Astronomy
SOHO	Solar and Heliospheric Observatory
SPADE	Stratospheric Photochemistry, Aerosol and Dynamics Expedition (NASA)
SSB	Space Studies Board
SSM/I	Special Sensor Microwave/Imager
SSP	space and solar physics
STAR	Science to Achieve Results (program)
STScI	Space Telescope Science Institute
TEM	transmission electron microscopy
TOGA	Tropical Ocean and Global Atmosphere (program)
TRACE	Transition Region and Coronal Explorer
UARS	Upper Atmosphere Research Satellite
UAV	uncrewed aerial vehicle
UNEX	University Explorer Program
URC	University Research Center
USDA	U.S. Department of Agriculture
USRA	Universities Space Research Association

D

Biographical Information for Task Group Members

Anthony W. England, *Chair*—Dr. England received his Ph.D. in geophysics from the Massachusetts Institute of Technology in 1970. His research interests have included terrestrial heat flow; geomagnetic and gravimetric studies in the Rocky Mountains and in Antarctica; radar studies of temperate and polar glaciers; and microwave radiometric studies of snow, ice, freezing soils, and planetary regoliths. He served as scientist-astronaut for the National Aeronautic and Space Administration's (NASA's) Manned Spacecraft Center from 1967 to 1972 and again as a senior scientist-astronaut from 1979 to 1988. He was mission scientist for Apollo 13 and 16, and he flew as a mission specialist on space shuttle Challenger's Spacelab 2, a solar astronomy and plasma physics mission, in 1985. He served as program scientist for the space station during 1986 and 1987. Between 1972 and 1979, he was a research geophysicist and the deputy chief of the Office of Geochemistry and Geophysics with the U.S. Geological Survey. Dr. England has been at the University of Michigan since 1988, where he is professor of electrical engineering and computer science; professor of atmospheric, oceanic, and space science; and associate dean of the H.H. Rackham School of Graduate Studies. His research group is developing land-atmosphere energy and moisture flux-radiobrightness models for prairie and arctic tundra. He has received several honors from NASA: the Outstanding Science Achievement Medal (1973), the Space Flight Medal (1985), and the Exceptional Achievement Medal (1988). Dr. England has also received the U.S. Antarctic Medal (1979) and the Flight Achievement Award from the American Institute of Aeronautics and Astronautics. Dr. England is a member of the American Geophysical Union and Sigma Xi, and a fellow of the Institute of Electrical and Electronic Engineers. He is also a member of the Space Studies Board.

James G. Anderson—Dr. Anderson received his Ph.D. in physics and astrogeophysics from the University of Colorado. His primary research interests are gas-phase kinetics of free radicals and photochemistry of planetary atmospheres; he was a pioneer in in situ detection of stratospheric free radicals from balloon and high-altitude aircraft platforms. Dr. Anderson was a postdoctoral fellow at the University of Pittsburgh from 1971 to 1972, and then a research assistant professor of physics from 1972 to 1975. From 1975 to 1978, he was a research scientist at the University of Michigan's Space Physics Research Laboratory; this was followed by a brief associate professorship with the Department of Chemistry and the Department of Atmospheric and Oceanic Science (April-July 1978). Dr. Anderson was the Robert P. Burden Professor of Atmospheric Chemistry at Harvard University from 1978 to 1982, and he is currently the Philip S. Weld Professor of Atmospheric Chemistry. Dr. Anderson is an elected member of the American Academy of Arts and Sciences (1985), a fellow of the American Association for the Advancement of Science (1986), a fellow of the American Geophysical Union (1989), and an elected member of the National Academy of Sciences (1992). He has also received the American Chemical Society National Award for Creative Advances in Environmental Science and Technology (1989), Harvard University's Ledley Prize for Most Valuable Contribution to Science by a Member of the Faculty (1989), the United Nations' Earth Day International Award (1992), the University of Washington Arts and Sciences' Distinguished Alumnus Achievement Award (1993), the Gustavus John Esselen Award for Chemistry in the Public Interest awarded by the Northeastern Section of the American Chemical Society (1993), and the E.O. Lawrence Award in Environmental Science and Technology (1993). Dr. Anderson was also a mission scientist for the NASA Airborne Arctic Stratospheric Experiment 11 (AASE 11) from 1991 to 1992.

Magnus Höök—Dr. Höök received his Ph.D. from the University of Uppsala, in Sweden, where he also worked, first as a teaching assistant (1971-1974) and then as assistant professor (1974-1979). In 1979, he was associate professor at the Swedish University of Agricultural Science. Dr. Höök taught at the University of Alabama at Birmingham from 1980 to 1992 in several capacities: associate professor and professor of biochemistry (1980-1992); professor of microbiology (1989-1992); associate professor of ophthalmology (1989-1992); and professor of cell biology (1989-1992). He holds several concurrent positions, which include director of the Helen Keller Eye Research Foundation (1988-present); director of the Center for Extracellular Matrix Biology at Texas A&M University's Institute of Biosciences and Technology (1992-present); professor of biochemistry and biophysics at Texas A&M (1992-present); adjunct professor of cell biology at Baylor College of Medicine (1993-present); adjunct professor of veterinary anatomy and public health at Texas A&M's College of Veterinary Medicine (1993-present); and adjunct professor of both ophthalmology and medicine at Baylor (1994-present); and he is the Neva and Wesley West Chair at Texas A&M's Institute of Biosciences and Technology. His primary research interests include extracellular matrix biology; molecular and cellular regulation of cell adhesion; bacterial interactions with extracellular matrix; and septic arthritis and pathobiology of cell adhesion. Dr. Höök received an American Heart Association Established Investigatorship Award in 1981-1986.

Juri Matisoo—Dr. Matisoo received his Ph.D. in electrical engineering from the University of Minnesota and is widely experienced in research management. Dr. Matisoo began his career in 1964 as a member of the research staff at the Research Division of the IBM T.J. Watson Research Center, specializing in low-temperature physics. In 1969, he moved to research management, holding a variety of positions, including director of silicon technology at the T.J. Watson Research Center, where he directed teams researching high-performance devices, process technologies, microprocessor designs, and a state-of-the-art pilot line for process testing, and culminating with a 7-year term as vice president and director of the IBM Almaden Research Center, IBM's facility for storage-related research. From 1981 to 1982, Dr. Matisoo served on IBM's Corporate Technology Committee, part of the Office of the IBM Chair responsible for providing technical advice and guidance. Dr. Matisoo retired from IBM in

1995 to serve briefly as the vice president of research for the National Semiconductor Corporation, where he was charged with developing a broad-based research function and directing its new National Semiconductor Research Laboratory (1995-1996). Dr. Matisoo currently works as a consultant. He is a fellow of the Institute of Electrical and Electronic Engineers (IEEE) and was recipient of the IEEE Jack A. Morton Award for outstanding contributions in solid-state devices.

Roberta Balstad Miller—Dr. Miller has worked and published extensively in the areas of science and technology policy and human interactions in global environmental change. Dr. Miller received her Ph.D. from the University of Minnesota. Formerly the president and chief executive officer of the Consortium for International Earth Science Information Network (CIESIN), she is now senior research scientist and director of CIESIN at its new home within Columbia University's Earth Science Institute. She was previously a staff associate with the Social Science Research Council (1975-1981), the founding executive director of the Consortium of Social Science Associations (1981-1984), and director of the Division of Social and Economic Science at the National Science Foundation (NSF; 1984-1993). She received NSF's Meritorious Service Award in 1993. Dr. Miller has served as chair of a number of scientific advisory groups, including the North Atlantic Treaty Organization's Advisory Panel on Advanced Science Institutes-Advanced Research Workshops; the Committee on Science, Engineering and Public Policy of the American Association for the Advancement of Science; the Human Dominated Systems Directorate of the U.S. Man in the Biosphere Program; and others. From 1992 to 1994, she served as vice president of the International Social Science Council. Dr. Miller is also a member of the Space Studies Board.

Douglas D. Osheroff—Dr. Osheroff, whose main research interests center around studies of quantum fluids and solids and glasses at ultralow temperatures, received his Ph.D. from Cornell University. He was a member of the technical staff at AT&T Bell Laboratories from 1972 to 1981, and served as head of the Solid State and Low-temperature Physics Research Department from 1981 to 1987. Dr. Osheroff began his tenure as professor of physics and applied physics at Stanford in 1987, and he is currently the J.G. Jackson and C.J. Wood Professor of Physics. He is a fellow of the American Physical Society, and an elected member of the American Association for the Advancement of Science, the American Academy of Arts and Sciences, and the National Academy of Sciences; he has served as vice-chair of the International Union of Pure and Applied Physics' Commission on Low-temperature Physics. Dr. Osheroff's honors include the Sir Francis Simon Memorial Prize (1976); the Oliver E. Buckley Condensed Matter Physics Prize (1981); the MacArthur Prize Fellow Award (1981); the Walter J. Gores Award for Excellence in Teaching, Stanford University (1991); and the Nobel Prize for Physics (1996) for his discovery, with David Lee and Robert Richardson, that the helium-3 isotope can be made superfluid at a temperature only about two-thousandths of a degree above absolute zero.

Christopher T. Russell—Dr. Russell received his Ph.D. in space physics from the University of California at Los Angeles (UCLA). The principal focus of his research is the energy flow from the Sun through the solar wind and into the terrestrial and planetary magnetosphere, both intrinsic and induced, and how this energy is dissipated within these magnetospheres. Other interests include the generation of the intrinsic magnetic fields of the Earth and planets, and nature and strength of planetary lightning. He has spent his entire professional career at UCLA, first as a research geophysicist (1968-1981), and then as a professor at the Institute for Geophysics and Planetary Physics (1982 to the present). His Space Physics Group at the institute is composed of an engineering team that builds spaceflight instrumentation; a data processing team that processes returned data; and a scientific analysis team of students, researchers, and fellow faculty. Dr. Russell has been involved with the spaceflight program for many years: He was a co-investigator on the OGO-5 fluxgate magnetometer, and a co-investigator on the Apollo 15 and 16 subsatellite magnetometer. He was the principal investigator for the International Sun-Earth Explorer 1 and 2 magnetometer and for the Pioneer Venus Orbiter magnetometer. Currently, he is a co-investigator on Galileo-IDS, a team member of the FAST polar Earth orbiter, principal investigator for the POLAR polar Earth orbiter, and a team member of the Near-Earth Asteroid Rendezvous (NEAR). Dr. Russell is also co-investigator for the Cassini Saturn orbiter launched in October 1997. He has received the Macelwane Award presented by the American Geophysical Union, and he is a fellow of the American Geophysical Union, American Association for the Advancement of Science, and Royal Astronomical Society. Dr. Russell has authored and coauthored more than 850 articles in scientific journals and books on topics in planetary and space physics.

Steven W. Squyres—Dr. Squyres received his Ph.D. from Cornell University. His major research interests are in the geophysics and geochemistry of Mars, the geophysics and tectonics of icy satellites, the photometric properties of planetary surfaces, the tectonics of Venus, and planetary gamma-ray spectroscopy. Prior to receiving his Ph.D., Dr. Squyres was a teaching assistant for the Department of Geological Sciences at Cornell (1977-1978) and a geologist for the U.S. Geological Survey in its Flagstaff, Arizona, Astrogeology Branch (1980). He was a National Research Council postdoctoral research associate at NASA Ames Research Center from 1981 to 1983, and a research scientist from 1983 to 1986. Dr. Squyres was an assistant professor at Cornell's Department of Astronomy in 1986; he also was a visiting assistant professor for the California Institute of Technology's Division of Geological and Planetary Sciences (1986) and a visiting associate professor for the Department of Earth and Space Sciences at UCLA (1986). He was associate professor at Cornell from 1989 to 1995, and has been professor from 1989 to the present. In addition, Dr. Squyres was an associate of the Voyager imaging science team from 1978 to 1981. He was also a radar investigator on the Magellan mission, a member of the Mars Observer gamma-ray spectrometer flight investigation team, and a co-investigator on the Russian Mars 1996 mission. He is currently a member of the imaging science team on the Cassini mission and a member of the gamma-ray/x-ray spectrometer team on NASA's NEAR mission. Dr. Squyres has received the Cornell University Department of Geological Sciences Buchanan Award, a National Science Foundation graduate fellowship, a Cornell University Andrew Dickson White Fellowship, the Antarctic Service Medal, the American Astronomical Society Harold C. Urey Prize, and two NASA Group Achievement Awards (1982, 1984).

Paul G. Steffes—Dr. Steffes received his Ph.D. in electrical engineering from Stanford University; his primary research area is microwave and millimeter-wave remote sensing of planetary atmospheres, microwave and millimeter-wave satellite communications systems, radio and radar astronomy systems and techniques, and noninvasive monitoring of glucose levels in the human body through stimulated raman emission. He worked as a graduate research assistant at the Massachusetts Institute of Technology's Research Laboratory of Electronics, Radio Astronomy, and Remote Sensing while pursuing his masters (1976-1977). From 1977 to 1982, he was a member of the technical staff at Watkins-Johnson Company's Sensor Development in San Jose. He was a graduate research assistant at Stanford University's Center for Radar Astronomy while pursuing his Ph.D. (1979-1982). Dr. Steffes has worked at the Georgia Institute of Technology since 1982, as assistant professor (1982-1888), associate professor (1988-1994), and professor (1994-present). He has been involved with several space missions, including Pioneer-Venus, Magellan, and the Advanced Communications Technology Satellite. He was a member of NASA's Search for Extraterrestrial Intelligence (SETI) Microwave Observing

Team and is currently involved with the Project Phoenix microwave search conducted by the SETI Institute. Dr. Steffes' honors include the Metro Atlanta Young Engineer of the Year Award, presented by the Society of Professional Engineers (1985); the Sigma Xi Young Faculty Research Award (1988); elected membership to the Electromagnetics Academy (1990); the Sigma Xi Best Faculty Paper Award (1991); NASA Group Achievement Award for the High Resolution Microwave Survey Project, for which he was principal investigator (1993); and the IEEE Judith A. Resnik Award (1996).

June M. Thormodsgard—Ms. Thormodsgard received her B.S. in mathematics and computer science from the University of South Dakota and her M.S. in environmental engineering from the University of Wisconsin. Ms. Thormodsgard worked as a mathematician at the Naval Research Laboratory's Space Science Division (1973-1978), where she conducted research on satellite radar altimeters and passive microwave instruments, as well as the operation of aircraft instrumentation for field verification. In 1978, she joined the Earth Resources Observation Systems Data Center and is currently the branch chief for science and applications, where she coordinates a multidisciplinary team (70 staff scientists and 20 visiting scientists) whose mission is to facilitate the use of remote sensing and related geographic information by Earth scientists and resource managers. The activities of the team serve as a bridge between academic, government, and private research institutions and the practical and operational programs for improved Earth resource mapping, monitoring, modeling, and management. Ms. Thormodsgard has received the Superior Service Award and Meritorious Service Award from the Department of the Interior.

Eugene H. Trinh—Dr. Trinh received his Ph.D. in applied physics from Yale University; his primary research areas are physical acoustics, fluid dynamics, materials studies, and technology development for NASA spaceflight experiments. He was a postdoctoral fellow at Yale from 1978 to 1979 and at the Jet Propulsion Laboratory (JPL) from 1979 to 1980, after which he became a member of the technical staff (1980-1985). Dr. Trinh was the technical group leader from 1985 to 1988, and then a project scientist and payload specialist astronaut from 1988 to 1992. He was involved in the development of shuttle flight experiments and participated in Spacelab flight mission support activities and flight crew training. He was the alternate payload specialist on the Spacelab 3 mission (1985) and research task manager and project scientist for the Drop Physics Module flight experiments. Dr. Trinh participated in the development and operation of low-gravity experimental apparatuses for tests in the NASA KC-135 airplane. He was a member of the crew of space shuttle Columbia for the STS-50/U.S. Microgravity Laboratory-1 Spacelab mission. Dr. Trinh is currently a senior research scientist and technical group supervisor at JPL and an adjunct professor for the Department of Mechanical Engineering at the University of Southern California. Dr. Trinh has received several honors from NASA: the Group Achievement Award for flight experiments; the Science Achievement Award for principal investigator team; the Exceptional Scientific Achievement Medal; and the NASA Flight Medal. Dr. Trinh was recipient of a Sheffield fellowship from Yale University.

Arthur B.C. Walker, Jr.—Dr. Walker received his Ph.D. in physics from the University of Illinois at Urbana. His research interests are focused on the development of innovative space-borne instruments for the study of high-temperature astrophysical plasmas and on the use of x-ray far-ultraviolet and extreme-ultraviolet techniques to study other astrophysical phenomena, such as elemental abundances in the interstellar medium. He began his professional career as a member of the technical staff at Aerospace Corporation's Space Physics Laboratory (1965-1968). He spent several years as a member of the Space Astronomy Project, first as staff scientist (1968-1970), then as a senior staff scientist (1970-1972), and finally as director (1972-1973). Dr. Walker has been professor of applied physics at Stanford since 1974 and he was associate dean of graduate studies from 1976 to 1980. Dr. Walker's group is currently focused on the study of the physical processes underlying the structure and dynamical behavior of the solar corona and chromosphere, using observations from a variety of spacecraft, including the joint NASA-European Space Agency Solar and Heliospheric Observatory (SOHO) and his group's Multispectral Solar Telescope Array. Dr. Walker's group is also establishing a Stanford Advanced X-ray Astrophysics Facility (AXAF) Science Center, in preparation for the launch of NASA's AXAF X-ray Observatory in 1998, which will provide access to AXAF observations for astronomers in the western United States. He is a member of Sigma Xi, the American Physical Society, American Geophysical Union, American Astronomical Society, and International Astronomical Union. Dr. Walker is also a member of the Hansen Experimental Physics Laboratory and the Center for Space Science and Astrophysics.

Patrick John Webber—Dr. Webber received his Ph.D. in plant ecology from Queens University in Canada. His research interests are broad, ranging from classical phytosociology and plant taxonomy to arctic ecology and the ecology of managed landscapes, which are prevalent in the U.S. Midwest; his current research focus is on landscape ecology and the evolution of managed and natural ecosystems. Prior to receiving his Ph.D., he was an assistant professor at York University's (Canada) Department of Biology (1966-1969). Between 1969 and 1989, he progressed from assistant to full professor at the University of Colorado's Department of Environmental, Population, and Organismic Biology. Dr. Webber was also director of the University of Colorado's Institute of Arctic and Alpine Research (1979-1986). From 1987 to 1993, he was program director for ecology at the National Science Foundation (NSF). Dr. Webber's first position at Michigan State University was as the director of the W.K. Kellogg Biological Station (1990-1993). He then returned to NSF as the program director for arctic system science. He has been professor of forestry at Michigan State University since 1990, and professor of botany and plant pathology since 1993. Dr. Webber has directed several large research projects, including the San Juan Ecology Project (1970-1976), the U.S. Alpine Program of the International Tundra Biome Program (1971-1974); he was the founding principal investigator of NSF's Alpine Long-term Ecological Research Program (1980-1987). He is currently principal investigator of an NSF 5-year award to study the effect of climate warming on tundra vegetation under the International Tundra Experiment project.

Ronald M. Konkel, *Consultant*—Mr. Konkel, who served as consultant to the Task Group on Research and Analysis Programs, received his M.A. in economics from Tulane University. He entered government service as a management intern at the NASA Johnson Space Center (1964-1966). He was a program analyst in the Office of Manned Space Flight at NASA Headquarters and later served as a program analyst and staff economist in the Office of the Comptroller (1966-1972). He served as a budget examiner and later as deputy division chief for the Energy and Science Division and as chief of the Science and Space Programs Branch at the Office of Management and Budget (Executive Office of the President, 1972-1980). He was staff director for the U.S. Senate Subcommittee on Science, Technology, and Space (1980-1981) and a senior economist in the Planning Office at the National Institute of Standards and Technology (1981-1982). From 1982 to 1989, Mr. Konkel served at NASA Headquarters as director of the Administration and Resources Management Division in the Office of Space Science and Applications. He spent a sabbatical as a visiting fellow at the Center for Space and Geosciences Policy at the University of Colorado (1988-1989). Mr. Konkel retired in late 1990 to work as a consultant and has completed assignments for NASA, the Universities Space Research Association, the congressional Office of Technology Assessment, and three organizations within the National Research Council.